Hans von Ohain: Elegance in Flight

The young inventor Hans von Ohain is lifted to the shoulders of his colleagues on 27 August 1939, celebrating the world's first flight propelled by a turbojet engine.

Hans von Ohain: Elegance in Flight

By Margaret Conner

American Institute of Aeronautics and Astronautics
1801 Alexander Bell Drive
Reston, VA 20191–4344

Publishers since 1930

American Institute of Aeronautics and Astronautics, Inc., Reston, Virginia

2 3 4 5

Library of Congress Cataloging-in-Publication Data

Conner, Margaret
Hans von Ohain : elegance in flight / Margaret Conner
p. cm.
Includes bibliographical references and index.
ISBN 1-56347-520-0 (alk. paper)
1. Ohain, Hans von, 1911–1998. 2. Aeronautical engineers—Germany—Biography.
3. Aircraft gas-turbines—History.
I. Title.
TL540.033 C66 2002 629.13`0092—dc21 2001056722

Cover design by Gayle Machey

Foreword

Just a little more than 60 years ago, the first turbojet flight took place on 27 August 1939 at Marienehe, Germany. Credit for inventing the turbojet engine is shared by Hans von Ohain and Englishman Sir Frank Whittle, whose turbojet engine powered an aircraft in 1941. In many instances through the years, Whittle has been incorrectly recognized as the sole inventor. In 1992, von Ohain and Whittle were awarded the prestigious Draper Prize for "early jet development and contributions to mankind."

No book about von Ohain's life and work has been published. Although reports and papers have been written by von Ohain and others, there has been no complete story until now. This book describes Hans von Ohain, a brilliant physicist who, his whole life, worked under the principle of solving difficult problems with simple solutions. We follow him from childhood through his education, the first turbojet development, and his work at the Heinkel Company. We see his dream of "elegance in flight" come true with the flight of the Heinkel He 178 aircraft powered by the turbojet engine he created. This book gives information about the "jet race" in Germany during World War II.

Von Ohain's story continues after the war, when he was blessed to come to the United States. There he met his wife and started his family. We read of his career with the U.S. Air Force, where he rose to be one of the top scientists in the field of advanced propulsion. We meet family members and friends, and share their memories and experiences with von Ohain.

This book has come into being because the U.S. Air Force wanted to recognize this great man who was formerly chief scientist of the Aero Propulsion Laboratory at Wright–Patterson Air Force Base, Ohio. Contributions to make it possible came from the aircraft engine industry. Information has been gathered from collections of writings, interviews, and site visits. Support and help in research, and for reviewing the material, came from people both within and outside the United States. The von Ohain family provided information and many pictures from the family collection. Peggy Conner at Universal Technology Corporation has done an excellent job of writing.

The book is a historical document, but it is also evidence of a man's dream of a lifetime coming true in the creation of "elegance in flight." What it has meant to mankind!

Erik Prisell
May 2000
Chief Engineer and Strategic Specialist for Aero Propulsion,
Swedish Defense Materiel Administration (FMV)

Author's note: Erik Prisell has made unique contributions to this book. He is an expert in propulsion in his own career and was a close friend of Hans von Ohain for many years. He is now a member of the von Ohain family as husband to von Ohain's daughter, Stephanie.

Contents

Preface

Beginning with my very first interviews, I learned from Hans von Ohain's friends and former associates that his real genius was in living, and his technical achievements were secondary to what he was as a person. Each person I talked with began with lengthy examples of what a wonderful person he was. Meetings and conversations with him were described as "uplifting." They said he listened and considered what was said, and gently offered suggestions that were usually brilliant insights. One friend said, "When he spoke with you, his eyes lit up, and he focused entirely on the discussion."

All of those interviewed said that von Ohain was exciting to work with because he was always looking to the future, to the next possibility, and shunned talking about what he had contributed in the past. Of his intelligence, one man summed up the descriptions with the word "enlightenment," a spiritual understanding beyond ordinary genius.

Those who knew Hans von Ohain wanted him to be remembered. I have made the story of his life from official documents, miscellaneous writings, and reminiscences of many of his friends. I apologize for the many people I did not contact. I may have missed some interesting and important stories. Most of the names cited should be credited with doctoral degrees, but I decided early that the story became cluttered by adding "Dr." to every name, and, in all cases, those interviewed spoke as Hans's friends, not merely professional associates. I hope they will forgive me for not giving full recognition to their status.

Vast amounts of information has to be laid aside. The book could not include more than passing references to World War II and the personalities involved and tells little of the enormous scope of research of the U.S. Air Force during the years from 1947 to the present.

The U.S. Air Force and Universal Technology Corporation made resources available for me to write this book.

My personal acquaintance with von Ohain made me the logical person to write this volume. I had known Hans von Ohain socially because his wife Hanny and I had the same voice teacher, Professor Theodor Heimann. She and I sang duets together for many years. I had worked in the Propulsion System Program Office at Wright–Patterson Air Force Base, Ohio. When I wrote a short presentation about von Ohain for the dedication of our conference room to his memory, many people who had known him approved of it. I was an experienced researcher, both in my profession as contract negotiator and as speaker for Dayton Opera lectures.

I thank the following for financial support for writing of the book: U.S. Air Force and Propulsion Directorate at Wright–Patterson Air Force Base, Pratt & Whitney, Rolls-Royce Allison, Honeywell Aerospace Engines and Systems, and Universal Technology Corporation.

Many individuals provided help with research. Kervyn Mach translated endless pages of German documents from the Heinkel archives and files of the Deutsches Museum, Munich. Marvin Stibich of the Air Force Research Laboratory, Propulsion Directorate, at Wright–Patterson Air Force Base provided a vast collection of material about von Ohain that had been saved in that office. Erik Prisell in Sweden, a propulsion expert in his own career and friend and son-in-law of Hans von Ohain, was a great asset because he could meticulously review all technical information with knowledge from his conversations with von Ohain, as well as provide family information. Volker Koos of Rostock, Germany, gave vital details from his research on von Ohain's work. Squire Brown, who is a student of the history of powered flight, helped guide me to a correct perspective of some events.

The von Ohain family gave me both encouragement and contributions: my dear friend Hanny von Ohain; the children, Steve Brown, Chris von Ohain, Cathy Pravel, Stephanie Prisell and her husband Erik; and Hans von Ohain's brother, Wolf von Pabst.

My gratitude goes to the following for their many hours spent in interviews, and the stories and documents they provided: Jim Aldridge, Max Bentele, Philippe Bouchard, Alice Bagnell, Hollis Conner, Ken Cramer, Tom Curran, Tony Fiore, Alan Garscadden, Sigmund Harder, Bill Heiser, Bill Jackomis, Ben Koff, Gunter Kruse, Tatjana Mauch, John Minardi, Jim Nelson, Paul Nikolai, Frank Patella, Jim Petty, Marv Schmidt, Leonard Spialter, Margaret Tucker, Henning von Gierke, Dieter Walk, Art Wennerstrom, and Dale Whitford.

Many people read the document during its writing. I thank them for corrections, questions, and encouragement as the pages grew: Erik Prisell, Bob Henderson, Howard Bethel, Tom Curran, Cathie Stropki, Bill McCuddy, Pam Kearney, Squire Brown, Volker Koos, Jim Petty, Janice Mach, and members of the von Ohain family.

The von Ohain quotes are from his writings, the record of the von Ohain/Whittle conversation, and the cited formal interviews. Sources for other quotes are described within the text.

This book can be a complement to the biographies of Sir Frank Whittle who worked in Great Britain to develop a turbojet engine during the same years von Ohain worked in Germany.

Margaret Conner
December 2001

Introduction

Of all the memorable flights since Kitty Hawk, the one on 27 August 1939 most directly links us to modern aviation. On that day the first aircraft powered by a jet engine made its initial flight. The engine was the design of a young physicist, Hans Joachim Pabst von Ohain, who, through remarkable insight and vision, succeeded where others had stumbled and gone astray. His vision was an "elegant" propulsion device, and his insight guided him along a path that turned the vision to reality.

The jet engine revolutionized aviation. It transformed the airplane from an expensive, special-purpose instrument into a reliable, affordable vehicle that could span vast distances at high speed. When young Hans von Ohain experienced his first airplane flight, the aircraft engine was just becoming sufficiently reliable for that new enterprise, the airline, to offer scheduled service. However, the benefit was confined to the few individuals who could afford the expensive tickets for short, overland routes, and continuous maintenance and frequent engine overhauls were standard practice for the airlines. Today, because of the exceptional reliability and long life of the jet engine, airlines can offer low-cost, intercontinental service using twin-engine transports, even over long ocean routes. Both the global economy and individual daily life have been improved.

The recent generation of historians of technology cites the importance of culture, environment, and institutions, as well as technical knowledge, for achieving success in inventions. Frank Whittle in Great Britain and Hans von Ohain in Germany made almost simultaneous, yet independent, inventions of the jet engine. Would the invention have emerged from impersonal aeronautical institutions, or did it require inspired work by especially gifted individuals? Was the device simply the next step in the deterministic advance of thermodynamic technology, or was it the outcome of the creative processes of these unusual men? Hans von Ohain was educated in the science of physics, not engineering, which allowed him to see opportunities in aircraft propulsion that eluded traditional powerplant specialists. Furthermore, he was a student at Göttingen University in Germany, the home of Professor Ludwig Prandtl and the birthplace of the scientific foundation of aeronautics. The culture of Göttingen helped shape the young graduate student's approach to the concept of jet propulsion. Invention of the jet engine required a competence in scientific, analytical disciplines that were seldom needed in the empirical development practices of reciprocating engine manufacturing.

The product of von Ohain's work, a practical turbojet engine, opened the door to flight at speeds much faster than the conventional reciprocating engine/propeller combination. Soon the sound barrier loomed as the next big hurdle. Military services in the principal industrialized nations realized the turbojet's potential advantages and began to invest in improvements that would increase its power, efficiency, and reliability. The results were high-performance military machines and engineering knowledge to transfer to civil, commercial products. Manufacturers of military jet engines began to find customers for airline applications, and airlines found the operating costs of jet aircraft to be significantly lower than their piston-engine types. Ticket prices fell, setting off the dramatic growth in personal travel that transformed our paradigm of distance and time.

At the end of World War II, von Ohain was a rather anonymous person, lacking the high-profile name recognition that might have been his if he had traded on his early successes. Many scientists in Nazi Germany chose to accept generous financial rewards and renown for their personal accomplishments in military technology, and emerged with reputations forever tainted. The principled character of Hans von Ohain is revealed in his continual service to his patron, Ernst Heinkel, and his loyalty to his own beliefs.

In the post-war period, von Ohain entered the classic American experience of the immigrant starting a new life. The U.S. Army Air Force offered him the opportunity for a second career, a second chance to pursue "elegance" in his propulsion research, at Wright Field (later Wright–Patterson Air Force Base). His invention of the jet engine was yesterday's task; the excitement of new problems stimulated the quest for new solutions. He became part of Air Force initiatives that launched a remarkable era for the acquisition of knowledge.

The exploitation of science and technology from the work of research scientists, including the Paperclip scientists brought from

Germany, created the strength for the free world to win the Cold War and brought us to the threshold of vast new knowledge. Hans von Ohain's invention of the jet engine and continued contributions to research won the respect and admiration of colleagues and friends. He left an exceptional legacy.

Squire L. Brown, Ph.D.
July 2000
Chief, Technical Assessment Division, Wright Laboratory,
Wright–Patterson Air Force Base, Ohio (Retired)
Member, Society for History of Technology

Acknowledgments

We gratefully acknowledge the following sources for providing illustrations and photographs included in this book.

National Air and Space Museum, Smithsonian Institution, p. ii
Georg August University, Göttingen, p. 27 (top right)
Deutsches Museum, Munich, pp. 129 (bottom), 131, 132, 133 (bottom)
ASME International, p. 271
Frank Woesthoff, pp. 27 (bottom), 228
Volker Koos, pp. 127, 128
Karl–Ernst Heinkel, p. 126 (bottom)
Max Bentele, p. 130 (bottom)
Hans Amtmann, p. 221 (top)

All other illustrations and photographs were obtained from the von Ohain family's collection.

About the subtitle

Hans von Ohain was 20 years old when he took a flight in a commercial aircraft. As a sports car enthusiast and a student glider pilot, he was shocked by the noise and vibration of the engine. He said, "The smoothness and elegance of flying was completely destroyed by the shaking, vibrating, noisy, stinking reciprocating engine." With his knowledge of thermodynamics, he set out to make a propulsion system that would be smooth, light, and powerful. In speeches later in life, he described his goal for flight to be "elegance."

Foreign-born von Ohain had to learn English and carefully selected words to express what he meant. "Elegance" is a puzzling term to describe flight unless you consult a dictionary that gives several definitions of the word. The *American Heritage Dictionary* gives not only "refinement and grace in movement," but another definition, "scientific exactness and precision."

Von Ohain achieved elegance in his work. His engine made flight more enjoyable. He was always exact and precise in his remarkable, creative solutions to scientific problems. Those that knew him say "elegance" also describes his dealings with people. He treated everyone with refinement and grace, while answering their needs with careful thought.

Von Ohain

Hans von Ohain was not particular about how you pronounced his name, and often called himself "von O-haayn" as a joke. However, if you wish to pronounce his name correctly, remember that in German the "i," being the second vowel, is the one to pronounce, "von Oha_i_n," emphasis on the "i." Or simply think, "von Ohine."

He asked his friends and those he worked with to call him simply "Hans," and he was "Hans" to everyone—with full recognition for his history and his professional position.

CHAPTER 1

Hans von Ohain: a description

Who has not stood on the edge of a hill looking out on the panorama below—villages, roads, pastures, great distances—and felt the air around you. You lift your arms and the wind catches your sleeves and your hair. You feel you can rise, can fly, like the birds around you floating quietly and effortlessly, higher and higher over the top of the world. Then you realize that your feet are planted on the Earth and even the grass around them is more free to move with the air than you are.

Humans have longed to fly since the beginning of time. Gliders were built to catch the wind, and the title "soaring" was attached to the sport. Humans tried to control and extend their flight, and a noisy little engine was added to the glider's frame. Soon larger and larger engines made more power but more noise for aircraft.

There was a young man who looked at the clouds one sky-blue day in autumn and said, "It must be wonderful to fly!" He was only a student, 20 years old.

He took a trip on a commercial plane with a reciprocating engine. He disliked the airplane's noisy vibrations and its smell of oil, and set out to make a change. He thought there could be a simple solution.

He designed an engine that worked as simply as a full balloon released to spin around a room. As with a balloon, a jet of air came out the back of his vehicle, hurling it forward with incredible speed and hardly a vibration, and with only a following whistle.

He called his engine's flight "elegant."

We now travel with quiet speed to the ends of the Earth, to places that before we might only read about. We can go tomorrow, funds and airline schedules permitting, and arrive almost the same day. With eager anticipation, we board our airliner. We buckle our seat belts as the doors are pulled closed. In a moment there is an almost imperceptible creeping from the gate across long aisles of concrete, and we pause. There is a roar for testing, a few more feet of creeping, a turn, and a new roar, more intense. The wheels grind down the runway. And then, lift! We lean back and are released into the air. We are flung higher and higher, above the pastures and hills, past weather and clouds, up closer to the sun. In near silence, we are resting on the air, higher than our earthly dreams of flying. In hours we are in a different place, sharing a different life.

We thank this man with dreams of elegance in flying. We thank his imagination and his ability polished by education. Hans von Ohain was the man who said, "It must be wonderful to fly!" and made it wonderful.

Hans von Ohain had two noteworthy careers in his lifetime. In Germany he was the inventor of the first turbojet aircraft engine that flew, and he developed the engine for the world's first jet fighter plane. He came to the United States in 1947 and rose to the position of chief scientist for research in propulsion for the U.S. Air Force, with the additional duty of advisor for national and international technical studies.

He was intellectually brilliant and had earned a doctorate in physics from one of the world's most prestigious universities, Georg August University at Göttingen, Germany. Beyond these credentials, he was a loving husband and father and a warm and humble person. Those who knew him and worked with him remember him fondly for his many positive qualities.

Von Ohain was a practical man. While his early goal was to be a physics professor at Göttingen, life took a turn for him. He created a new form of propulsion, and in his lifetime he had the extremely rare opportunity to guide the technology from invention to its universal application. In later years he mused, "These 'children' that you put into the world, like [the jet engine], force you to stick to them longer than you really intended. My plans were to go back to the University. I was always intensely interested in the fundamental sciences as well as applications. But I didn't want to go back as an unsuccessful man. I said I would stay (with Ernst Heinkel) until I had a turbojet in flight. I didn't know that this was going to be a switch for life." Von Ohain, with his usual sense of humor, loved to say, "When you are a young man, you look around at all the beautiful young girls and realize that some day you are going to have to settle for just one. Well, I fell in love with that girl, namely, the jet engine. Pretty soon I got married to the jet engine idea and had to kill off the many other loves

I had. My main love was physics, but that died out and became an amateur diversion."

Von Ohain was modest. He discounted his invention of the turbojet engine, saying merely that it was an idea whose time had come: "If Edison would have died just prior to making the record playing device, someone else would have invented it. If Whittle and I and some other people would have died prior to our having done our work on the turbojets, we would still have turbojets today. Nature plays the same safe game in invention as it does in biological development. Technological development follows a certain path." He acknowledged the fast pace of technology: "In our time the path of development is steeper with respect to time." Every lecture he gave on development of the turbojet began with praise for the work of Frank Whittle and sympathy for the lack of encouragement Whittle received. Von Ohain always cited the help he received from every coworker and wanted to share honors with them. In 1991, when von Ohain and Whittle jointly received the prestigious Draper Prize, "the Nobel Prize of engineering," von Ohain wrote to Ernst Heinkel's son in Germany to share the honor with those who had worked with him, saying:

> Many thanks for your very essential help for the success of the Heinkel jet engine development. I thank my coworkers, especially Gundermann and Hahn, and later Dr. Bentele; my professor Dr. R.W. Pohl; and Dr. Ernst Heinkel, who took me into his aircraft firm in 1936. Heinkel's contributions to jet development were the maximum, because no one else in Germany in 1936 took such a large gamble with his own money.
>
> In thankful remembrance, let us all rejoice in the award of the Draper Prize.

He brushed off compliments, and accepted trips to receive honors only if he could bring his wife along because she would enjoy the ceremonies and interesting locations. It was not until he was older that he commented that he missed many opportunities to promote his work, and maybe should have taken more credit for himself.

Von Ohain had charm in how he dealt with people. He was described as a many-faceted diamond because he could relate to each person and situation differently and individually, from broad knowledge and experience. He treated those he worked with, from top internationally acclaimed scientists to food servers behind a cafeteria line, with respect and, frequently, droll humor. He listened to and appreciated each one and showed his appreciation with his own offerings: technical advice and suggestions to the scientists, and compliments and candy to the food servers. Alan Garscadden of the U.S. Air Force Research Laboratory said, "He was a very famous person and chief scientist, but you were not apprehensive. He was not there to critique as much to encourage." Life was more pleasant after he had come by.

Tom Curran, who followed von Ohain as chief scientist at the U.S. Air Force Propulsion Laboratory, described him as "insightful and incisive." One awesome assessment of his ability was, "He knew how nature worked." He could listen to an advanced scientific idea, hear it explained, think and analyze, and know if it could be made to work. He weighed alternatives, saw possibilities, and took risks. With his strong academic background and years of experimentation on his own projects, von Ohain had a sense of the direction research should take. He managed to keep a universal look at future needs for the Air Force in spite of the pressure of everyday business. Through interviews with his former coworkers, there is a consensus that they felt privileged to have known such a great genius.

Those who knew him well saw that he was an intensely private man who cherished the time he could spend with his thoughts. When he retired from Wright–Patterson he said, "Now I will have time to think," but his work was what filled his mind. He made contributions to scientific knowledge up to almost his last days.

A physicist studies the nature of matter and *why* things work. Hans von Ohain was a well-schooled theoretical physicist. He taught himself to be an engineer. His invention of the turbojet aircraft engine was a demonstration of genius rare in technical history. Von Ohain was not from the aviation industry and had never piloted an airplane. The engine was based on pure theory. Curiously, his family background was not in science and engineering, but in the military. However, his family fed his curiosity in all things, from poetry and literature to travel and natural science, which isolated itself into his love of pure science. Von Ohain's concept of an engine that had no moving parts had to be abandoned for a more immediately practicable plan, although in the last years of his life he may have achieved his original idea. Von Ohain, the theoretical scientist, met Ernst Heinkel, the wealthy businessman, and together they reached a zenith in human technology—flight of a turbojet-powered aircraft in 1939.

An engineer studies *how* things work. In England, Whittle was a pilot in the British Royal Air Force. The idea for his turbojet was evolutionary, derived from knowledge of the requirements of powered flight that he had learned from flying aircraft with reciprocating engines. His invention idea received a patent in 1930. His subsequent education at Cambridge University provided him with a sound practical and theoretical training in mechanical science and aeronautical engineering, mathematics, and physics. The engine he developed in the following years was entirely different in configuration from von Ohain's, but with the same basic concept. Because of the cost of such an advanced and unproven concept, Whittle struggled to receive financial backing to build a prototype to fly. Its first flight was in 1941.

Von Ohain was fortunate to connect early in his career with an entrepreneur who financed development of his invention. Although Whittle steadfastly continued to believe in his concept, he met with discouragement and frustration in attempts to obtain funding from both the British government and industry for his turbojet engine until World War II's necessities opened the potential for this new weapon. Thirty years after World War II, von Ohain and Whittle met in the United States, became close friends, and had long talks about the difficulties of the development of their engines and the world's reaction to the coming of the new form of propulsion. "If you had been given the money," von Ohain said to Whittle, "you would have been six years ahead of us. If Hitler or Göring would have heard that there is a man in England who flies 500 mph in a small experimental plane, and that it was coming into development, it is likely that World War II would not have come into being."

Von Ohain did not receive attention and credit for creating the first jet engine to fly until late in his second career, while Whittle was toasted and feted for his jet engine throughout his life. Perhaps the news of von Ohain's flight did not reach the world press because of other events in Germany. Perhaps Whittle received sole credit for the invention for many years because Whittle was on the winning side of World War II and von Ohain on the losing side. Inventor Samuel Pierpont Langley crashed his experimental aircraft into the Potomac River near Washington, D.C., on 8 December 1903. Nine days later, two brothers named Wright had a successful controlled flight on the sand flats of North Carolina. Samuel Pierpont Langley was given credit for the first invention "capable of sustained, man-carrying free flight" by the venerable Smithsonian Institution in the

United States, until 1942 when the Smithsonian finally acknowledged that Orville and Wilbur Wright were first in flight. Can anyone recall who was second in flight after the Wright brothers?

We remember von Ohain for the creation of the turbojet engine that had the first flight, and see the remarkable contributions to propulsion technology he made for 50 years more. Through the eyes of his family and friends, we learn to know the man.

Chapter 2

Family and education

Hans-Joachim Pabst von Ohain grew up in an upper-class home with the privileges of a comfortable, large house, servants to care for the children, holidays at the seacoast, and the expectations that he would grow into a professional career.

He was born on 14 December 1911 in Dessau, Germany, into a family distinguished by military honors. The von Ohain family was nobility, an award bestowed as a result of significant acts performed by a member of the Pabst family in the past. The *Heine Chronicle* of 1604 recorded the Pabst coat of arms, a golden shield bearing three golden six-pointed stars on a blue fess. Those initiated into nobility had the responsibility to uphold the highest standards for their families and to pass those standards to succeeding generations. The children were expected to be successful and to display the best manners and behavior. Honor and respect for others were of primary importance.

Hans von Ohain's family could trace their lineage back to the Middle Ages. The family name was originally "de Pape" when they lived in the town of Ohain in Belgium near Brussels. Ohain is remembered in history as the staging area for the Battle of Waterloo. When ancestor Paul de Pape was killed at the battle of Liege in 1468, the family fled to Saxony and changed their name to the Germanic "Pabst." The relocation brought good fortune to the family because this area was the center of mining of silver, copper, tin, precious stones, and salt.

The titles of "Baron" and "von Ohain" were added to their family name in the 1700s when Gottfried von Pabst, a student of metallurgy and geology, along with his friend Baron von Tschirnhaus, developed the first glazed white porcelain in Europe, Meissen china, named after the town where it was made. It was first produced in 1713, replacing porcelain previously available only in the Far East.

Gottfried von Pabst's son Johann was the first of the von Ohain family to be in the military. He became a cadet in the army of Saxony when he was 14 and saw action in Poland, the Balkans, Hungary, and Alsace. He rose to the rank of colonel as commander of a regiment.

Gottfried von Pabst's grandson led forces against Russia in two battles and became part of Napoleon's Grand Armee. He lost his life in 1812 at the Beresina River when Napoleon's forces tried to cross out of Russia and were completely destroyed. Another of Gottfried's grandsons was captain of a ship on the Great Lakes in America. There he fell in love and married the daughter of a Milwaukee brewer, Philip Best. In 1883, he became president of the brewery, which by this time had become the largest in America, the Pabst Brewing Company.

Rudolph Pabst von Ohain, great-grandson of Gottfried, was born in 1846. His career in the German military awarded him the rank of lieutenant general, comparable to the rank of a four-star general in the U.S. Army. He was thereafter addressed as "His Excellency." His sons were Wolf and Walter.

Wolf Pabst von Ohain, Hans von Ohain's father, was born in Naumburg in 1881 and sent at age nine for the strict military education at the Corps of Cadets in Potsdam, Germany. Of slight stature, he distinguished himself as a gymnast and horseman.

In 1906 in Magdeburg Cathedral, he married the beautiful Erna Nagel from a family of wealthy farmers of the Boerde region. To marry, the young couple had to receive permission from Wolf's commanding officer. This was a good marriage between wealth and aristocracy. Tragically, Erna died in childbirth, leaving a son, Herbert. In deep grief, Wolf left his son with grandparents and accepted a transfer for two years to the German East Africa colony near Lake Victoria. Wolf was an adventurer, going for months on safaris where he was the only European for miles around. Hunting trophies decorated the von Ohain homes for many years and were especially remembered by the family at the Berlin house in suburban Dahlem in the 1930s.

Wolf Pabst von Ohain returned from Africa to be stationed at Dessau. In January 1911, the 30-year-old officer married Erna's 21-year-old sister Katharina Louise, called Kitty. She was remarkably beautiful and kept her beauty into advanced age. They were very different in personality. She loved nature and animals. Hans's father enjoyed social life such as balls and dances. The first son of Wolf and his second wife was Hans-Joachim Pabst von Ohain. He was born on 14 December 1911 into a life of formality and graciousness almost forgotten in our present day. The family was not rich but retained a prominent position in society.

The family lived modestly in Dessau as a military family until World War I drew Wolf away from home. His wife and young sons Herbert and Hans-Joachim moved to Magdeburg where they could live near the Nagel family for support and encouragement. Maj. Wolf Pabst von Ohain's battalion fought on the front lines of many battles. He received Germany's second highest decoration for valor, *Hohenzollern-Orden mit Schwertern*. His career nearly ended when a bullet pierced his neck. Miraculously, the wound was not fatal, but damaged his vocal chords temporarily. The family moved from Magdeburg back to Dessau on his return from military service.

Wolf Pabst von Ohain was away for long periods performing military service during the boys' early years. He was a strict, stern disciplinarian in the military tradition and made sure that the boys were trained in proper social conduct and decorum. He was less interested in their academic achievements than was their mother, who scolded them and offered them the pleasures of all kinds of books. For Hans, especially, there were science books given to him on his birthdays to excite his curiosity. The family had a nanny who cared for Hans and the new baby, Wolf, born in 1921. She composed poetry, which delighted the boys.

Hans took joy in being a big brother to baby Wolf. As they grew up, he made him layouts of railroads and models of airplanes copied from *Friedolien* magazine. These models were not available in kits, as today, but had to be cut and whittled by the maker from light woods such as balsa and bamboo. Propulsion was by rubber band. Little Wolf enjoyed the attention and remembers Hans's experiments with "shooting electricity" [sic], especially the time he burned a huge hole in the family's formal dining room table, much to the displeasure of his mother and the staff. Hans always wanted to know how things worked by trying them out himself.

Hans was curious about science, and his family and their friends challenged him to learn about it. One question puzzled him for years. He said, "When I was 10 or 11, the father of one of my friends was a physicist. When he visited me he would ask all kinds of funny questions. He asked me what would happen if we put a fly in a jar filled with air—would the weight of the jar change when the fly sat still versus when it flew around?"

Hans's father saw that business as a profession was replacing the military for social status as the Weimar culture was recovering from World War I and the postwar chaos. When Hans was 13 the family moved to an apartment in Berlin where Wolf, the father, became a partner in the Walter Meyer Company, a distributor of light bulbs. Wolf's persistence and fair dealings with customers expanded the business and made him financially successful. In 1928, Wolf Pabst von Ohain contracted to have a house built on Gehegestrasse in Dahlem, Berlin's most fashionable suburb, then a district of broad avenues and magnificent villas. Designed by Germany's most outstanding architect, Herrmann Muthesius, the house was the biggest and most extraordinary in that neighborhood even compared with other large, expensive houses around it. Walls were of dark red brick a full yard thick. It had four levels. The lowest included a kitchen with a walkout at a sunken level that provided a small yard for receiving supplies and for laundry. The main level with bay windows looking out over the street had formal rooms for entertaining. The level above it was the living space for the family. It was large enough for each child to have a suite. A full staff lived on the fourth floor consistent with the style of the era. The house had a very small surrounding yard. As more people moved to Dahlem in later years, houses were built very close next door.

For 16-year-old Hans, the suburb's greatest charm was its proximity to the Grünewald forest, only a five-minute bicycle ride away. He spent pleasant afternoons outdoors when the family thought he was doing his homework.

Hans enjoyed visits with his colorful, extroverted uncle, Walter Pabst von Ohain, who was quick to build his own house in Dahlem, again using the architect Muthesius. It was even bigger than his brother's house and had a big yard surrounding it. Uncle Walter outshone his stern brother Wolf by being chauffeured around in American cars, including a 16-cylinder Cadillac. Besides holding social gatherings at his Dahlem house, he leased Germany's biggest hunting preserve where he was host to kings and princes. Uncle Walter was the head of a large German corporation, Pintsch. He had married one of the richest young women in Berlin and had an extravagant lifestyle. Herbert, Hans, Wolf, and their parents took advantage of his hospitality frequently.

The family made sure that their children had many experiences. Young Hans particularly enjoyed family holidays at resorts on the Baltic Sea where he could swim and sail. During his university years he was required by the National Socialist government to have a course in physical training. Most of the training was provided by Hitler Youth organizations, but Hans did not want to participate in any politically oriented activity. He attended a summer yachting school at the Baltic Sea, with the fee paid for by his Grandmother Nagel. It fulfilled the government requirement for training, was not Party affiliated, and pleased Hans, who loved sailing.

The family provided Hans's brother Wolf the unusual opportunity at age 15 to go to England as an exchange student. Such an experience was exceptional during those early Nazi years because association with "foreigners" was discouraged and Party beliefs were promulgated in all the schools. Wolf flew alone from Berlin to Croyden Airport and stayed for a year with the family of a colonel at Bedford, England. The "exchange" of the English student living with the von Ohain family never took place because of political tensions in Germany. Wolf spent some of the following year studying in France. The family had a number of foreign visitors, including return visits from Wolf's English and French friends. Wolf's ability in languages was valuable for his employment after the end of World War II.

Hans attended the *Arndt-Gymnasium* in Dahlem. A *gymnasium* was a rigorous school of classical humanistic curriculum for students preparing for university study. The standard of teaching was high, representing scholarly culture as well as a strong scientific background. The studies were very formally presented with little opportunity for creative problem solving. A student who graduated from a classical *gymnasium* could be considered to have the equivalent of a college undergraduate associate degree in the United States. For students headed for nonacademic careers, technical schools provided some fundamentals of mathematics, physics, and chemistry, but no in-depth theory or background.

Hans's school performance was expected to live up to that of his older brother, Herbert, who was brilliant and excelled in all academic subjects. Herbert was also handsome and self-confident, a difficult kind of big brother. Herbert was studying law and became a very successful corporate lawyer. In contrast to Herbert, Hans was a mediocre student in the kind of study that required attendance at formal lectures and rote learning. Hans described studying Latin, Greek, and French as "torture" in the beginning, although he later mastered the material sufficiently to receive credit.

After years of indifferent school performance, in his later *gymnasium* years Hans studied Newton's mechanics. His teacher introduced the work of that radical new thinker, Albert Einstein, and compared Newton's thinking with Einstein's. Hans said:

> At that time, the most fascinating thing in physics was Einstein's theory of relativity, which was still so new it was not part of the curriculum. This teacher was an outstanding person. He created a course for three or four interested students and began to teach us Einstein's theory. Einstein had written a very beautiful booklet explaining the theory of relativity for people who had only a high school education. It's amazing that he did that. It was our textbook for this extracurricular course. The teacher explained how the theory resolved the inconsistencies in Newton's mechanics. He was also very good at evoking the underlying philosophy in math and physics.

Hans found inspiration from a quote from Einstein: "When I examine myself and my methods of thought, I come to the conclusion that the gift of imagination has meant more to me than any talent for abstract positive thinking."

Hans was exhorted to use his imagination and to dream of what might be possible. Up to this time learning had been a method of storing up information. Hans saw that Einstein had learned his lessons but went on to stretch his mind to the farthest reaches of possibility. Hans's grades underwent a dramatic change in response to the challenge of examining Einstein's practical processes of reasoning. He suddenly showed his instructors enormous improvement, especially in the disciplines of mathematics and science. The boy who always wanted to try things for himself learned the discipline of education. The pattern Hans followed all his life emerged: try new ideas, even very complex ones, and find a way to arrive at simple, practical solutions.

German schools did not test their students frequently, but rather relied on the results of the body of learning as tested in all subjects at the end of several years of study. Hans's responses to questions on the comprehensive examination at the end of *gymnasium* years were extraordinary. The examiners were shocked, believing that he had obtained copies of the questions in advance of the examination. They required him to be retested, and again his scores were very high. A grade report of 13 March 1930 shows the following:

ARNDT-GYMNASIUM
IN BERLIN-DAHLEM

ZEUGNIS DER REIFE
(Proof of Completion, i.e. Diploma)

Hans Joachim Pabst von Ohain

2 Jahre auf dem hiesigen Gymnasium und zwar 2 Jahre in Prima
(2 years in regular high school and 2 years in "sixth form" or senior high school)

Religion	gut
Deutsch (German)	genugend
Lateinisch (Latin)	genugend
Grieschisch (Greek)	genugend
Französisch (French)	genugend
Geschichte′(History)	genugend
Erdkunde (Geography)	gut
Mathematik (Math)	sehr gut
Physik (Physics)	sehr gut
Zeichnen (Mechanical Drawing)	genugend
Leibesubungen (Fitness exercise)	gut

Er hat die Reifeprufung bestanden. (He has passed his examination.)

H.J. Pabst von Ohain will Physik studieren. (He wants to study physics.)

(Translation of grades: Grades in a German school cannot be directly compared with the American school grading of As, Bs, and Cs. Grades were not given liberally. *Genugend,* meaning "satisfies requirements," was the standard grade given for an adequate job of learning the class material. *Gut,* which was a very rewarding "good," might be given if a student showed better knowledge of the subject. It was very seldom that a student received the grade *sehr gut,* which would be "outstanding.")

In the early 1930s attendance at universities and technical institutes in Germany was at an all-time peak. Germany was awakening to the importance of science and technical skills. In spite of generations of military careers in the von Ohain family, the decision was made for Hans to go to a university to study physics. His interest was in science, and his personality was peace loving. Hans's family selected the Georg August University at Göttingen, the most prestigious school in Germany and comparable to Harvard in the United States, with the same broad curriculum requiring the study of history, literature, and philosophy, as well as science. The university was founded by imperial decree in 1737 and from the beginning stressed academic excellence.

Acceptance to schools of higher education in Germany depended totally on the preparatory grades and promise shown by the student in undergraduate education. The prestige or wealth of the student's family could not admit a mediocre student. Hans's qualifications were unquestioned for attendance at Göttingen. Tuition was not required. The only costs were books and living arrangements.

In a sentiment not unusual for an 18 year old, Hans was eager to escape from the family and go to the university.

Hans von Ohain began studies at Georg August University in 1930. The school was dedicated to pure thinking in all academic disciplines, particularly natural science. Hans called Göttingen "the city of the pure scientific idea" in which the political realities of the 1930s "did not fit at all." Göttingen, referring to both the university and the town where the school is located, was a German Protestant college town that could be described as a cloistered, tranquil oasis of learning.

Göttingen today has probably changed little on its exterior from the time that Hans von Ohain attended the university there, or even 100 years before that time. The town was not bombed during World War II because the Allies wanted it untouched. The center of Göttingen is dominated by a medieval city hall and a large open square with a fountain topped by a statue of the *Ganseliesel* (the Little Goose Girl), a young girl tending her geese. She reminds the townsfolk of Göttingen that the Brothers Grimm wrote many of their fairy tales while they lived there. Narrow, twisting, cobbled streets lead to churches founded as early as the 10th century and to half-timbered houses that retain their faces, although they may have been rebuilt behind. Student gathering places, the *Zum Altdeutschen* and *Zum Szultenburger,* still welcome carefree celebrations. The charming old buildings and the interiors of the *Bierstuben* look like stage settings for a German operetta, and one expects to hear a student prince singing around the next corner. The school does not have a central campus. Instead, university buildings are scattered among the buildings of

the town itself, making it a true university town. The university's present enrollment of 33,000 students live throughout the town as they did in the past. Theodor von Kármán, the famous aerodynamicist, described the somber university buildings as having "scholarly elegance and culture." Tall brick buildings close together house laboratories and offices that have been upgraded with equipment but retain their aged dimensions. The vast school library has only recently been replaced with a glass and steel structure.

Göttingen retains an outstanding scholastic reputation internationally, offering classes in theology, law, politics, medicine, mathematics, philosophy, the arts and education, natural sciences, forestry, and agriculture. The lecture hall for physics has seen as many Nobel laureates as any hall in the world. The Max Planck Institute, formerly the Kaiser Wilhelm Institute and one of the leading science research organizations in the world, has its administrative headquarters at Göttingen. At the time Hans entered Göttingen University, the faculty included some of the world's top scientific intellectuals—Ludwig Prandtl, Albert Betz, and Walter Encke. Von Kármán was an occasional lecturer during von Ohain's student years.

Ludwig Prandtl can be described as the most important single contributor to the development of modern fluid mechanics. He taught at Göttingen from 1904–1946. He had already addressed compressible subsonic aerodynamics in the 1920s. He was named director of *Institut für Angewandte Mechanik* in 1925 and was founder of an aerodynamics experimental station at Göttingen. Prandtl helped in the development of Germany's highly sophisticated wind tunnels. During World War II he was advisor to the Air Ministry of Germany.

Albert Betz was director of the Institute of Aeronautical Research at Göttingen. His specialties were the theories of air and fluid flow.

Walter Encke was another expert in aerodynamics. He researched axial flow compressors. Later he presented papers on single-stage compressors and was a consultant to the jet engine development team at Junkers under the direction of Anselm Franz. In Hans's years at the Heinkel Company, Encke provided assistance as a consultant.

Von Kármán was Prandtl's most illustrious student. He would later be a major influence on Hans's career in the United States.

The physics lecture hall where Hans spent many hours had a large inscription in Latin above the blackboard and projection screen, *"Simplex sigillum veri"* (Simplicity is the sign of truth), an

adage that became etched into his memory. Once again Hans found that teaching was by the formal lecture method, but he adjusted to this form of instruction by hearing and taking notes to analyze and research further. Hans described the university years:

> Under the German educational system you can develop an inferiority complex because you don't have many exams. You have gigantic initial lectures in huge rooms with loudspeakers. You can hardly see the professor, and no one takes attendance. You don't pay much for lectures, but it's up to you to learn the material. You don't have an exam until much later, three semesters or more. So you have no feel at all for how you're doing. About three years later you have an extremely tough final exam that covers everything from Adam to Eve [sic]. I like the U.S. system better, where you have tasks and are given questions. At Göttingen we received a very, very fundamental education, and we enjoyed it. Even those who were application-oriented enjoyed it, even though they didn't hear anything at all about applications in the first eight semesters. But we got a coherent picture of science. You could simply put all the building blocks together, and they made sense. You didn't learn simply for passing exams, or you didn't lose yourself in too many specialties. You really got the most important fundamental laws. You learned to play with them. In mechanics, for instance, you learned about the First and Second Laws of Thermodynamics and their enormous application in science.

Hans had no problems with the initial course in physics, but his first class in mathematics was from a teacher who taught in the abstract, and Hans wanted something more. He repeated the course with the renowned Richard Courant, another of the university's professors. Hans said, "The difference was like night and day. Courant was oriented toward theoretical physics, and he was so merciful to the beginner. He explained everything using concrete examples, and he had so much humor. He was an amateur skier. I remember how he compared shock waves to skiers coming down a hill: 'When one hits a tree, the others follow and crash into him. That's how shock waves are formed.'"

Under Prandtl and Betz, Hans studied thermodynamics and aerodynamics, everything, he said:

> From incompressible flow to compressible flow to boundary layer to turbulence. I got a nice, clear picture of aero- and thermodynamics. It was really the applied mechanics of Newton, only much more sophisticated. Newton's ideas of how lift and drag occur were wrong. That's why the Wright Brothers had to go all the way back and solve that problem. They had to make experiments in their own wind tunnel, and later they had to learn about circulation around the wing. That's really what causes lift.

Under Robert Wichard Pohl he learned the foundations of modern solid-state physics. Pohl had an international reputation for teaching and demonstrating physical principles.

Students studying toward a doctorate in physics were invited to attend some lectures at the Kaiser Wilhelm Institute. Hans was fascinated by theories, and germinating in his mind was a way to put theories into practical use. It is apparent that Hans learned what the classes provided because he developed an outstanding technical background. He was enthusiastic about new ideas. In the evenings, when he gathered with other students at the Göttingen *Bierstuben,* there were heated discussions, and he challenged his friends to work with him on experiments.

Hans's first classes, physics, mathematics, and chemistry, from April 1930 to June 1931, were at Göttingen University. He continued the same studies in Berlin from September 1931 to January 1932, and studied physics, mathematics, and theoretical aeromechanics at Rostock University from April 1932 to July 1932. He returned to Göttingen University in September 1932.

Hans liked to try out the professors' ideas for himself, whether in calculations or in physically tinkering with apparatus to find a different or better way to work. His professors saw that he learned quickly and had a mind that saw beyond the knowledge that already existed. He became friends with Professor Prandtl, Professor Betz, and particularly Professor Pohl, his physics professor in his undergraduate years and, later, mentor for his doctoral studies. Pohl's particular interest, semiconductors, covered the whole area of photochemistry. Professor Pohl was the first to make large monocrystals, a very brilliant accomplishment in this area.

Hans explained:

> In Germany, the professor and student do not usually have a very close relationship. A student comes to the professor and

> says, "I would like to do my doctor's thesis with you." The professor is hesitant because he doesn't know whether the student is good, bad, or mediocre, and so he gives him little tasks to perform and thereby gets to know him better. It was in one of these tasks that I first demonstrated my flair for applications to Professor Pohl. He said to me, "You know, you could help me on a little task. I want [to have] prepared for my demonstrations to the medical people, and for many others, an apparatus to demonstrate the method of obtaining the refraction index of different substances in a simple way. All these instruments we have here work beautifully, but they are so complicated that a student never really understands them quickly. Now, what could you do there?"
>
> I said, "Let me think for a while and I will try to come up with something simple." It was quite a brain teaser for me. But after two weeks or three I came up with a model which really excited Professor Pohl very much because it was really the utmost in simplicity.

Hans chuckled at his clever solution. "It was ridiculously simple. Professor Pohl enjoyed the apparatus very much. After this, Professor Pohl asked me what I was interested in doing for my thesis."

Because Professor Pohl agreed to be the mentor for Hans's doctoral thesis, Hans wanted to select a subject relating to Pohl's field of expertise. At this time, Hans did not consider that his musings in aerodynamic theory relating to a propulsion device for flight would be a suitable topic. Pohl did not try to influence his student in the choice of a subject. Hans said:

> He left me pretty much alone, but we had nice discussions about physics and became better acquainted. He gave me one personal lecture that I remember well: "You know, Hans, the thing is this, either you do good physics or you do really creative applications. I would hate to see you, as my student, making nothing out of technology but doodle signs, so to speak, and nothing really creative." I told him about my ideas on the turbine for aircraft and he said, "When you have something which really revolutionizes technology by employing physics, this is fine, but don't do mediocre stuff." He encouraged me very much to go into the applications area because "You have a flair and a talent for that."

Von Ohain's doctoral thesis was titled "The Application of Zero Order Optical Interference for Translation of Sound Waves Directly into Electrical Impulses," or, more simply, "An Interference Light Relay for White Light." He called it an optical microphone. He remembered:

> I had an idea about how to transform sound waves directly into light vibrations. The sound movies were around the corner at the time, and I thought it would be very timely to get them a much better microphone with less distortion than the existing ones.
>
> The invention describes an optical microphone which operated by modulating a very thin film of a metallic mirror. This, in turn, modulated the light intensity, which then was recorded in a photo cell. With this optical microphone we could make excellent reproduction of sound of a quality comparable to that of the present capacitor microphone. But the capacitor microphone requires a very good amplifier. My idea was to eliminate the amplifier by having the sound directly translated into light impulses so that you could, for instance, record it on film. My thesis showed that pressure pulses could be accurately converted into light pulses on a linear scale with no distortions with respect to frequency, nor with respect to amplitude. The device produced no nonlinear and no linear distortions. It was absolutely true up to sound frequencies of 20,000 cycles per second sound waves, that means, to the upper acoustical limit with absolute fidelity from very low to very high frequencies.
>
> It's amazing: I could sell the idea very quickly. The microphone got an international patent. Siemens bought it for about 3,500 Marks. It was not much, but I enjoyed it very much. The optical microphone did not really enter technology. The Siemens Corporation was astonished by the fact that we got such enormously strong output impulses from a small device of about one cubic centimeter in volume. It seemed to Siemens to have great technological possibilities. The unfortunate thing was the fact that these little metal foils never could hold the adjustment for zero interference. It required adjustments on the order of a fraction of a micron. That turned out to be impractical; as a matter of fact, technically impossible. It was eventually discarded.

While Hans von Ohain was dedicated to learning, he also made sure to take time to enjoy life at the university. His memories included good times at student hangouts in the town. He ice skated and skied, sailed and swam. His brother Wolf reported that Hans took classes in ju-jitsu and liked to come home and demonstrate on his little brother. His family financed his books and rent for a comfortable flat, and provided a small amount of spending money. His family did not want him to be a "playboy," but Hans's Grandmother Nagel spoiled him by sending extra spending money and indulged him with special privileges his family did not think he needed.

In his first year at Göttingen, Hans was happy when his brother Herbert transferred ownership of his automobile to him. In Germany only 1 in 100 of the population owned an automobile, and it was most unusual for a student to have a car. It was an Opel, not a luxury car, but it impressed young women because it was sporty and its top folded down. Hans charmed them by giving driving lessons. His driving was reckless enough that his friends were convinced he was trying to kill them. The car gave him the chance to live in an apartment away from the main part of the campus, a route he remembered well, "toward Geismar." He had the freedom to go with his friends and a professor or two, including his good friend, Professor Karl Stellmacher, to ski in the nearby Harz Mountains.

Hans needed to maintain his car. In those days an automobile demanded constant maintenance. A car's owner usually became an experienced tinkerer in the various workings of the car, but it helped to have a long-time relationship with a trustworthy auto repairman. Just a few blocks away from the university's physics building, within walking distance, was the Bartels and Becker auto repair shop. It was a convenient place for von Ohain to leave his car for repair or maintenance in the morning on his way to class and pick it up at the end of the afternoon. (The Eden Hotel is now at that location.) Hans stopped in frequently to discuss his car with the chief mechanic, Max Hahn. The two became fast friends. Appropriate to the times, each addressed the other formally as "Herr Doktor von Ohain" and "Herr Hahn." This formality remained in their later professional careers.

For Hans, life was a pleasure, studying what he loved, living as he wanted. Hans wanted no commitment such as marriage. Professor Pohl had high hopes that young Hans would marry one of his two daughters, who were geniuses in mathematics. While they were intelligent and attractive, Hans was not interested.

Hans's brother Wolf describes him in those years as "rough, straightforward, not submissive," and not materialistic. Hans hated anything new and shiny. Wolf said that the family gave Hans new, expensive luggage for his birthday, but when he returned home for a visit a short time later, the luggage was scuffed and scratched. Hans had taken it outside and tossed it around in the dirt until it looked used.

Hans had a good wardrobe, but was sometimes distracted about what to wear. He appeared at his home one weekend in the dead of winter in a summer white linen suit. He liked the suit and knew he looked good in it. That was what mattered to him.

During Hans's university years of 1930–1935, critical changes occurred in Germany. In the early years economic depression was deep. Thousands were unemployed and many were starving. Hans was privileged to attend college. In 1933, when Hitler became leader of the country, his National Socialist Party was at first seen as a palatable alternative to communism and the creeping Soviet influence. After Hitler provided government programs to build highways and public buildings, employment and the standard of living improved. At the same time professors and friends of Jewish background disappeared one by one, either escaping from the country or being arrested and taken to unknown destinations. Hans's outstanding mathematics teacher, Richard Courant, was Jewish and was one of the first professors ousted from Göttingen by the Nazis. He had a distinguished career later as a professor at Columbia University. Nazi teachings were promoted and newspapers carried only selected stories and propaganda. Campus organizations were infiltrated with National Socialist representatives. Theoretical physics was probably the subject matter least affected by the new philosophies. Hans isolated himself from politics during his intense years of studies.

Hans von Ohain, with his dark hair and eyes, did not fulfill the image of Aryan looks that were considered the ideal German appearance. One day as he was walking home from school with his arms full of parcels, and deep in thought, a group of Nazis passed him with their flag. Hans's arm did not go up in the Nazi salute to the flag quickly enough for the group. The leader stopped and hit him and pushed him to the ground. "Jew!" he shouted.

Before long, Hans was certain that he had found what he wanted to be in life, and that was a full professor at Göttingen. He had a talent for explaining and enjoyed working with challenging questions that were brought to him. He knew his position had to be a full professorship. He explained:

> The universities in Germany pay fabulous salaries to the "king" professor, the top man, but very little to the small man, his assistant. The difference between the salaries of an assistant professor, an associate professor, or a full professor were far, far greater in Germany than in the United States. An assistant can hardly support a family with his salary. It was, however, an honor to be an assistant of a professor. If an assistant could do a phenomenal job, make discoveries, etc., he might become a full professor in twenty or thirty years and then get a good salary.

To remain at the university and keep his comfortable bachelor apartment, he had to be financially independent. Hans planned to achieve this goal by becoming an assistant professor, augmenting the assistant's income by patenting something and selling the patent to some company. Perhaps if he invented something unique he could earn the position of full professor at an early age. The optical microphone had sold quickly. He would have to think of another invention.

The oral examination for Hans von Ohain's doctorate was conducted by Professors Pohl, Sauter, Prandtl, and Herglotz.

Prandtl commented, "Grasp of subject good, breadth of knowledge not large. Grade: Good." Hans was unhappy with these comments. He said:

> Some weeks before my examination in applied mechanics, I had said to Professor Prandtl that I was excellent in aerodynamics and thermodynamics, but not in materials science. There was no class in that subject in the short time available. I thought that Professor Prandtl would not include materials science in my examination. Unfortunately, he made the test almost entirely on the subject.
>
> Professor Pohl and Dr. Sauter tested me in the subject of physics. I knew my way around this subject very well, but had trouble with Dr. Sauter's questions as being unclear. I discovered later that Professor Pohl had similar difficulty with them.

Although grades for his answers varied from "satisfactory" through "very good" Hans was concerned about his inadequacies. Those who have endured the oral examinations for a doctorate can be sympathetic.

Professor Pohl's evaluation of his dissertation shows his appreciation for his pupil:

> The light relay constructed by Mr. von Ohain has meaning, not only for the technical purpose specified in the work, but is also adaptable for application in many physical tasks. Mr. von Ohain conceived the idea for this design entirely on his own and carried out the physical investigation of the relay's characteristics with a refreshing self-reliance. With the completion of the work, not only has he pulled off a great experimental undertaking, but also demonstrated the ease with which he could apply his knowledge of theoretical physics to practical problems. I can without reservation recommend a grade of "Very Good" for Mr. von Ohain's work.

Pohl's words, "refreshing self-reliance," are an apt description of von Ohain's approach to most of the work in his life. He was willing to take the responsibility to work out solutions to problems, whether technical or relating to management. The usual seven-year course of study to complete a doctorate was accomplished by von Ohain in four years. On 1 November 1935 he received the diploma for his doctorate in physics with minors in aerodynamics, aeromechanics, and mathematics.

Without a doubt, Hans von Ohain would not have left Göttingen without joining in the tradition of every doctoral graduate. According to custom a crowd follows the new graduates to the statue of the *Ganseliesel* (the Little Goose Girl), which stands two tiers up on a pedestal in front of the huge medieval *Rathaus* (city hall). The symbol of the town of Göttingen, her figure with her geese stands smiling with demure grace above the fountain in the center of the square. Every new doctor makes the giant step to the first level of the fountain followed by another giant step across the pool up to the second level to reach up and kiss the Little Goose Girl amid the cheers and merriment of his fellows. The crowd is likely to visit one of the *Bierstuben* close to the town square both before and after the ceremony. Von Ohain laughed to tell about kissing the Little Goose Girl. When his good friend, Sig Harder, was climbing up to kiss her, he missed his step and fell into the fountain. Sig was finally able to kiss her on his second attempt, "but her lips were very cold."

Von Ohain's parents asked him what he wanted as a graduation gift and, characteristically, he did not want a material gift. He asked for a trip to northern Italy, to Lake Garda. He and his brother Herbert spent days sailing and enjoying the beauty of the landscape. Von Ohain said, "This is something I can keep forever in my mind's eye."

Years later when he took his wife, Hanny, to vacation in the same area, he reminisced about how quickly his life changed in the years after this carefree vacation.

Maj. Wolf Pabst von Ohain with his one-year-old son, Hans, and Katherina Louise (Kitty), Hans's mother, with Herbert.

Baby Hans.

Wolf and Kitty Pabst von Ohain with their sons Herbert (left) and Hans (right) skiiing in the Harz Mountains.

From left, Hans, Wolf Jr., and Wolf Sr.

Grandmother Nagel with 16-year-old Hans flying a kite while on vacation at the Baltic Sea.

Hans's first car, photographed in front of the house in Dahlem, near Berlin. This car was given to him during his college years by his brother Herbert.

Logo of Georg-August University, Göttingen.

The town square in Göttingen with the figure of Ganseliesel (the Little Goose Girl) on top of the fountain.

Unter der Amtsführung
Seiner Magnifizenz des Rektors Professor Dr. phil. Friedrich Neumann
ernennt
die Mathematisch-Naturwissenschaftliche Fakultät der Georg August-Universität
durch ihren Dekan Professor Dr. phil. Jonas Schmidt

Herrn Hans Joachim Pabst von Ohain aus Dessau

zum Doktor der Philosophie.

Die von ihm vorgelegte Dissertation:
»Ein Interferenzlichtrelais für weißes Licht«
wurde mit
»Sehr gut«,
das Ergebnis der in Physik, Theoretischer Physik, Mathematik
und Angewandter Mechanik
am 16. Februar 1935 abgelegten mündlichen Prüfung mit
»Gut«
beurteilt.

Zum Zeugnis dessen ist diese Urkunde ausgestellt und mit dem Insiegel der Fakultät versehen.

Göttingen, den 1. November 1935.

Der Dekan
der Mathematisch-Naturwissenschaftlichen Fakultät

Doctoral diploma for Hans von Ohain, awarded 1 November 1935. He received a "very good" for his dissertation and a "good" for studies in physics, theoretical physics, mathematics, and practical mechanics.

Hans von Ohain, age 24.

Chapter 3

Idea for a propulsion system

Hans von Ohain showed brilliant intellectual ability from his college preparatory work, through college, and on to studies for a doctorate. He enjoyed learning. Von Ohain was unusual because he had an interest beyond the single path to a doctorate. While he was researching and preparing a thesis on one subject, the optical microphone, he was spending serious time calculating a hypothesis for a new form of propulsion.

When von Ohain heard lectures of famed Ludwig Prandtl on the aerodynamics of gliding, he joined the university soar plane (glider) group, the Academic Soar Plane Sports Club. He found gliding to be a beautiful outdoor sport. He was active in the club through his university years until it became politicized by the National Socialist organization and renamed the National Socialist Flying Club. He resigned when they did more marching and talking than flying. Hans wanted no part of the political movements that were creeping into many campus organizations. As a result of membership in the gliding club, Hans decided to try a flight in a powered aircraft.

Von Ohain had watched the airliners of the 1920s and 1930s float serenely through the ocean of air above. He was 20 years old when he made a commercial flight from Cologne to Berlin on a Junkers Trimotor aircraft. "I was a physicist, and my interests weren't really in flying or in airplanes, and I was not too knowledgeable in this field, but somehow it was such a nice autumn for the gliders I said, 'Well, it must be wonderful to fly!'" Von Ohain visibly shuddered when he recalled his first ride:

> The propellers made a horrendous noise. The airplane rattled because it had piston engines. The elegance of flying was totally taken away by the reciprocating ch-ch-ch and vibratory thing, by the shaking, vibrating, noisy, stinking, reciprocating engine.

> You couldn't even talk to your neighbor. It was difficult to believe you were flying at all unless you looked out of the window. It was not as romantic as I thought it would be. So I got interested and said, "Well, gee, that isn't good." As an uninhibited thinking physicist you come up with another process.

Von Ohain was indeed uninhibited, or, more accurately, unknowing. He had never been a pilot or an engine mechanic and was unfettered from the history of previous ideas to use jet propulsion for aircraft. He had theories derived from his education. Von Ohain's goal was to find a way to make flying as beautiful as gliding, as he said, "like a fantastically fast glider." He wanted "elegance" in flying. In lectures that he made throughout his life he chose to use the word "elegance." It is interesting to compare von Ohain's goal with that of Britain's Frank Whittle, who shares the credit for design of the first turbojet engine. In 1930, with the filing of an application for his first patent, Whittle said he believed "that an embodiment of this invention will provide a large thrust in proportion to its weight, that it will perform at greater altitudes than are at present obtainable, that it makes possible higher speeds than have up to the present have been obtained, that it will operate with any fuel now in use, and that it will have a reasonably low fuel consumption. Further, that simplicity and convenient external form is achieved." Whittle envisioned that faster planes could carry mail with more speed and might fly mail back and forth across oceans. But Whittle did not express the dreamer's goal of beauty in flight.

Von Ohain pursued a hobby of aircraft propulsion, determined to get some special attention at Göttingen. He explained:

> My primary interest was in the natural sciences with aeronautics as more of a hobby. I wanted to work on a project where I could put my knowledge in both fields into play. I had a strong flair for applications. It was a flair for the unconventional type of application that relied on the use of physics to do certain novel things. My physics background was very sound. I understood thermodynamics, classical physics, and modern physics very well. But, of course, in "modern physics" in my time [his university years] they had not yet discovered the neutron. That came a year after I received my doctorate. Atomic physics was in its childhood.

"When you are a young physics student," von Ohain said later, "you have intellectual arrogance; you think you can do anything." Hans made some casual sketches on paper for a propulsion device with the simplest possible structure and the lowest functional risk.

Von Ohain explained how his ideas began: "My interest in jet propulsion began in the fall of 1933 when I was in my seventh semester at Göttingen University. At that time the reciprocating engine with variable-pitch propeller represented an excellent propulsion system. No immediate need for new propulsion systems seemed to exist." The traditional engine manufacturers were still making improvements to the piston engine. It had been increased in power, but also greatly increased in its mechanical complexity, weight, size, maintenance requirements, fuel consumption, and cost. The airscrew (propeller), which the engine drove, had its own inherent complexity and limitations. Its efficiency dropped when its tip velocity reached the speed of sound. The airscrew also added weight to the aircraft.

The design began in his head. He thought it through as Professor Prandtl had advised was the first step in a process. He reviewed his theory with analysis and calculations. He proposed that the exhaust itself should be the driving force of propulsion rather than its driving a generator or a propeller. He acknowledged, "I didn't know that many people before me had the same thought." He believed that when better powerplants were available, airframe designs could be capable of performance far in excess of anything yet flying. In order to avoid high-speed compressibility problems, the propeller would have to be replaced by something else in order to achieve the speed of sound.

His first plan was a configuration with no moving parts:

> My thoughts began to focus on a steady aerodynamic flow process in which the energy for compressing the fresh air would be extracted from the combustion gas without machinery. Such a steady flow process promised a far greater air-volume handling capability than that of a reciprocating engine, and, consequently, a greater power concentration and power-to-weight ratio. Also, the air ducted into such a system could be decelerated prior to reaching any Mach-number-sensitive engine component. Both of these characteristics are of greatest significance for a high-speed propulsion system. It could be inherently more powerful, light, smooth, and compatible with future high-speed

air vehicles than a propeller-piston engine. I intended to accomplish this process without employing moving machinery by bringing the in-flowing fresh air in direct contact with the expanding combustion gas, a kind of ejector process.

But after studying specific processes and configurations, it became apparent that such types of processes would have enormous problems with respect to internal losses and adverse heat transfer effects caused by mixing between fresh air and combustor gas. I put this idea aside for future consideration and began to investigate a propulsion process in which compression and expansion were separated, and carried out by a turbojet compressor and turbine, respectively.

Von Ohain began again with a new set of calculations:

Searching for an extremely lightweight, compact, and simple configuration having a minimum development risk, I decided to work on a gas turbine process which could be called a modified "Nernst" turbine having a radial outflow compressor rotor back-to-back with a radial inflow turbine rotor. This configuration also promised correct matching simply by providing equal outer diameters for the straight radial outflow compressor rotor and the straight radial inflow turbine rotor. I was aware of the possibility of employing axial flow compressors and turbines, and I considered an axial flow configuration as very desirable for future developments from a standpoint of small frontal area, but as too complex and expensive for the beginning. In particular, stage matching of a multistage axial flow compressor, and matching of axial flow compressor and turbine without component test facilities, appeared to me too risky.

He worked intensively to validate the calculations with his knowledge of physics. He related his continuing thought process:

During 1934, I conducted rudimentary design and weight studies and made some performance calculations based on a pressure ratio of 3:1, which appeared attainable with a single-stage compressor and a turbine inlet temperature of about 1200 to 1400 degrees F. (650–750 degrees C.) It appeared that at a high flight speed of about 500 miles per hour (800 km/h), an overall efficiency could be obtained which would be around 60

> percent [compared with] that of an equivalent propeller engine. The corresponding high fuel consumption was somewhat discouraging. However, the weight of such a propulsion system promised to be only a fraction, a quarter or less, of that of an equivalent propeller-piston engine system. At that time the propulsion system of a fighter aircraft constituted a much greater weight portion than the fuel. Consequently, the above trade between fuel weight and propulsion system weight seemed to be a very favorable one.

Von Ohain's early calculations did not show that his concept of the turbojet engine might have greater efficiency than reciprocating engines of the time, and high fuel consumption was a concern that would receive negative comments from critics of his idea.

Von Ohain thought of his initial efforts for the turbojet as a "device," not an engine. "I thought an 'engine' was something where you grease surfaces, and where pistons are reciprocating and crankshafts are rotating," he said. "I thought it can hardly be considered an 'engine' except for its having bearings. But I wasn't an engineer. It was mostly sheet metal, and the bearings were really incidental." As to the steam turbine, he said, "This to me was something which I would say was already a monster of a machine. All heavy things. Grates and boilers. I considered that the turbojet was essentially a rotating thin sheet-metal unclassifiable device."

Von Ohain wanted to patent his turbojet idea. He was already familiar with patent procedures because he had patented the optical microphone. He discussed his ideas with a patent attorney in Berlin, Ernst Wiegand, who was the first to call the turbojet the "jet engine." The patent search showed that there was quite a bit in the field of jet propulsion already, although not in the form of the turbojet. When Wiegand made a search he found Paul Schmidt's patent on the pulse jet and Lorin's on the ramjet but nothing on the gas turbine-jet. Von Ohain remembered applying for about two or three patents. The first was entitled, *"Verfahren zum Umwandeln von Warmeenergie in kinetische Energie eines Gasstroms"* (Process for Converting Heat Energy into Kinetic Energy of a Gas Stream). These patents were the foundation for the later contract with Ernst Heinkel, although they had not been awarded at the time von Ohain went to work for Heinkel.

Von Ohain asked the patent office whether his design should be considered "secret." At first, they replied that they saw no reason to classify it that way, but they asked that von Ohain let them know about the device after it was built. Von Ohain's Patent Number 317/38 was finally validated as of 10 November 1935, and marked "Secret."

The patent attorney happened upon a Whittle patent during the years that the von Ohain patents were being formulated. This was the second Whittle patent; Whittle had not renewed the first one because he could not afford the fee. About Whittle's patent von Ohain said, "We felt it looked like a patent of an idea. You find many idea patents. For an invention, it is not necessary that you build it. We thought it was not being seriously worked on. I had not the slightest idea that Whittle was at that time working very keenly along similar lines but with different technical approaches." Hans described the difference:

> Our structure was very unconventional and completely different from Whittle's in almost every respect. Our patent claims did have to be narrowed in comparison to Whittle's, because Whittle's showed certain things. When I saw Whittle's patent, I was almost convinced that it had something to do with boundary-layer suction combinations. It had a two-flow, dual entrance flow, radial flow compressor that looked monstrous from an engine point of view. Its flow reversal looked to us to be an undesirable thing, but it turned out that it wasn't so bad after all, though it gave some minor instability problems.
>
> The patent attorney found a jet engine patent by A. Lysholm (of the Swedish company, *Milo Actiebolaget*). Sweden had a large number of proposals for beautiful gas turbines and turbojets using radial flow compressors. They looked very good to us, but these were patents on paper. No actual work had been done on these ideas. Actually, turbojet patents of the Frenchman Guillaume (French physicist, Charles Edouard Guillaume) should have been the cause for the rejection of practically all further turbojet patents. Why the German patent examiner didn't find Guillaume's patent is nebulous to me. I learned that the English patent examiner didn't find it either, so he granted Whittle a patent far too broad, which should not have been granted to Whittle.

Many years later when he was chief scientist at the U.S. Air Force's Aerospace Research Laboratory, von Ohain was asked if he had done a literature search into the subject of turbojet propulsion at the time he was pursuing his new idea. "No," he answered. "I have usually followed the routine of looking into what exists only after I had formed a certain thinking of my own. It is better to rely on your own thinking at the outset and reinvent everything for yourself because with this method you are likely to avoid the errors hampering other thinkers." Because von Ohain had been trained as a physicist, he was not hampered by what engineering schools consider to be "good engineering practice," that is, 1) search of literature and experience for what has been done previously, 2) conceptual incorporation of the new features, and 3) selection of the plan most likely to work. As leader of the laboratory, he said, "I encourage my co-workers very openly when we have to go into a new area not to start with a literature survey. You are too prone to follow the thinking of what already existed and obviously did not come to a successful end."

Theodor von Kármán, the influential aerodynamicist, also promoted this kind of thinking and said, "Exaggerated concern about what others are doing can be foolish. It can paralyze effort and stifle a good idea. One finds that in the history of science almost every problem has been worked on by somebody else. This should not discourage anyone from pursuing his own path."

Von Ohain urged his students:

> You have to have something which shows a clear departure from present technologies. When you have reached this advanced point in your thinking, then you are ready to look into the literature. Eventually you have to find out what other people did in this field. But when you do it too early, you are discouraged. When you have established hopes and avenues, it might be time to look into past literature. You can dampen enthusiasm very quickly. If I would have looked through the literature on the gas turbine, including the patents, I would have found that there was really nothing new, and my enthusiasm would have been dampened. The turbojet was therefore not a complete novelty. Since I was not familiar with its past history, it was a novelty for me. It helped me not to know what already had been done.
>
> The Frenchman Guillaume, as I learned much, much later, obtained a patent (international) in 1921, long before anybody else. (It was) on the axial flow compressor, a combustor, and on the turbine to drive the compressor and an exhaust nozzle very

> much like that of Lorin's ramjet invention of 1913. (He had) already invented the axial flow turbojet engine! Good that I didn't know!

In a lecture he gave in his later life, von Ohain said with a smile, "You always have to look at what the French have done. Historically, they seem to have an idea 20 years before anyone else, as Lorin and Guillaume in this case."

Because von Ohain had not researched previous literature, he did not know of the many negative opinions concerning turbojet propulsion. For example, in 1920 W. J. Stern of the British Air Ministry made a pronouncement: "In its present stage of development the internal combustion turbine is unsuitable for aircraft on account of weight and fuel consumption, the main difficulty in the case of the aircraft being the design of a light, compact, and efficient compressor." Stern based his decision on his own cast iron compressor, which was 6 ft long, 3 ft in diameter, and weighed 2000 lbs.

Before Whittle and von Ohain, other scientists had worked on turbojet development. In Great Britain, an important loser in the turbojet race was Alan Arnold Griffiths. He published a paper in July 1926 entitled "An Aerodynamic Theory of Turbine Design." Griffiths saw his engine as a compressor and turbine working together to drive a conventional aeroplane propeller. It was to be designed on the axial flow principle, in which air is forced from the front to the back of the engine through row upon row of blades making up the compressor and the turbine that drives it. Griffiths failed to comprehend that the jet flow itself was all that was needed to propel the airplane. A paper by Griffiths in 1930, reviewed by the British Aeronautical Research Committee (ARC), gave important findings indicating that the turbine engine would be "superior to existing engines in every respect." The ARC recommended that problems should be investigated, but, unbelievably, nothing happened! Griffiths's ideas languished. He might have designed the world's first working jet engine had it not been for the serious economic depression in Britain in the late 1920s, and his intellect, which was more academic than practical.

What is painful is Griffiths's treatment of Frank Whittle when they first met in 1930 and in the years following when Griffiths held an executive position with the British Air Ministry. His influence was a strong reason for Whittle's frustrating delays.

Von Kármán blamed himself for the decision of the U.S. military in 1938 to lay aside the idea of development of jet propulsion. In his book, *The Wind and Beyond,* he explained that he was on a U.S. Navy committee looking into the possibility of the development of turbojet technology. Von Kármán was out of the country when the Navy voted. Without reading the committee notes in detail, von Kármán agreed in absentia that jet propulsion should not be pursued. He later stated that one must always read the proceedings of a committee carefully and make one's own decisions.

Von Ohain was certain that his idea for turbojet propulsion was practical. He had spent more than a year of thought and analysis on paper and decided that the best demonstration of his theories could be made with a model. He said "the purpose of the model was to demonstrate the principle of stagnation pressure combustion in conjunction with the modified Nernst turbine." He was uncertain about how much the model would cost, and he counted what he had available. His pay as a teaching assistant for Professor Pohl amounted to very little. He had some money saved from his allowance and some money advanced by his father. Now that he had completed work on his doctoral thesis he could sell some of his optical instruments. He had money from selling his microphone patent. In the past, his maternal grandmother had given him funds for special projects such as his month of yachting school, and he turned to her again, this time to support his propulsion idea. In later years, a headline in the German press said, *"Die Oma finanzierte den ersten Jet"* (Grandma Financed the First Turbojet Engine).

To build the model, von Ohain went to his friend Max Hahn at the Bartels and Becker repair shop and garage, where he had his little sports car serviced. The late afternoon was a comfortable time for von Ohain and auto mechanic Hahn to get into discussions of cars. They had little in common except their interest in how things worked. Hahn was a dour man in poor health, 10 years older than von Ohain. He had been an apprentice on the German railways before working in the garage. Although Hahn had no formal education in physics or technical subjects, it was apparent to von Ohain that Hahn was an outstanding natural engineer and an expert machinist. Von Ohain liked to hear Hahn relate experiences from his considerable knowledge of manufacturing methods. Von Ohain showed Hahn his calculations and preliminary sketches of the turbojet device. Hahn's education was at the practical level, and the volume of calculations on paper meant little, but as he examined the simple drawings and listened to von Ohain's explanations he was confident that he could put together this young student's device. His taciturn, gloomy attitude balanced the enthusiasm of the young student.

Hahn was both amused and amazed at young Hans von Ohain. He seemed to be only a boy with the enthusiasm of a teenager. He knew very little about how to make something out of metal parts but had good ideas drawn on paper. Hahn would have to explain the process of making something and show him how it was done. Young Hans didn't mind getting his hands into the process even if it involved pinched or cut fingers, or dirt and oil. What amazed Hahn was the boy's response when something didn't work: He would laugh and come up with a change, and the change was always an improvement. There seemed to be no end to the ideas coming from that young mind. Hahn hadn't done anything he enjoyed so much as working with this young man. He looked forward to every afternoon when they could get together.

"Herr Doktor von Ohain" and "Herr Hahn" continued their discussions of the turbojet apparatus, details of what materials could be used, and how the pieces could be put together. The cost of a demonstration model was worked out and agreed upon. Other mechanics in the shop agreed to assist them if adequately paid. Hahn was honest and straightforward about the prospects of success with the device. He was willing to build it, but he said, "It'll never work." As von Ohain said later when they were planning to take the mechanism to demonstrate to Ernst Heinkel, "He was right. It didn't." As they worked together, the combination of the mechanic's practical mind and von Ohain's ingenuity made for an ideal collaboration of experience and ideas. Hahn adapted the design with his knowledge of construction methods.

Von Ohain said that in a little storage room behind the shop, "We made a little model according to my own sketches. It was about three feet in diameter and about one foot wide. The blades were made of heavy sheet metal. Almost nothing was welded because Hahn was more a man for clean machining operations, and for fastening with bolts and nuts." It was a pancake engine for either vertical or horizontal thrust, a very light design. It was little more than bent sheet metal. Von Ohain called it "a working model born of optimism." When asked if he considered using ready-made components such as an impeller of a water pump for the compressor, von Ohain replied, "I wasn't familiar enough with such things. My lack of technical knowledge saved me from doing anything of that nature."

When it came time to test the model, von Ohain turned for help to his friend Professor Pohl at the university. Pohl was fascinated with his student's leisure time activity. The professor was very open-minded and positive about all his students' experiments. Because von Ohain was a graduate student at the Physics Institute, he was unhesitatingly loaned instrumentation, a starter motor, and other equipment for the trials of the model. Essential measurements of temperature and pressure distribution were taken. Professor Pohl authorized that experiments with the device could be conducted in the courtyard of the physics building. The physics laboratory was a low wooden structure standing within a courtyard about 8 yards wide by 20 yards long and surrounded by four-storied dark red brick lecture halls and administration buildings, one of which housed Pohl's office. It was easy to slide the "device" out of the laboratory and into the courtyard for testing. There were a number of practice runs with roaring exhaust flames almost reaching the building at the end of the courtyard. When the engine blew up, Professor Pohl insisted that further experiments should be permanently relocated. He could not risk the possibility that von Ohain would burn down the academic buildings.

Von Ohain moved his device to the back room of the Bartels and Becker garage. It was exciting to von Ohain and Hahn if the device worked at all. Each time the model was tested, long yellow flames streaked out. Von Ohain later said, "It looked more like a flamethrower than a gas turbine." Von Ohain knew little about combustion. It obstinately refused to take place in the chambers designed for it but did in the radial turbine rotor extending into the exhaust jet. The malfunctioning of the combustors was substantiated by temperature indications on the metal surfaces and the formation of lampblack deposits. Max Hahn, normally very stern and skeptical, seemed in this instance to be quite positive and optimistic. He conceded, in a rare glimpse of a sense of humor, "At least the flames came out of the right place," and with high velocity. The engine did not self-sustain but did result in the unloading of the starter. Instead of the pounding of the conventional engine there was a smooth flow of power, and a new sound, a piercing whistle.

Von Ohain soon realized that he had spent all his money—his earnings, his savings, money from the patent, money from his optical equipment, and money he had begged from his grandmother.

Max Hahn's practical experience and suggestions had kept the model's construction cost as low as possible, about $250 (in U.S. dollars), which was, von Ohain observed, "quite a bit of money for a student in those days, but for a brand new invention, chicken feed." For purposes of comparison, this amount was approximately the cost of a medium-priced used automobile in 1935.

Von Ohain said, "Tests indicated to me that fundamental combustor investigations and systematic developments were necessary, but would require time and money exceeding my private means." He further explained, "It was not possible for me to build a model with high tangential velocity which could demonstrate performance capability. I could afford only disks and blades of constant-thickness sheet metal. The maximum tangential velocity was set at 150 m/sec. At high tangential velocities, the blades would have been flying about our ears!" Frustrated, von Ohain took his idea and computations to Professor Pohl, who was mildly amused at the desperation of the young student and encouraged him to get a proper theoretical analysis of the model down on paper.

Pohl gave him new courage. In an interview later in his life, von Ohain told what Professor Pohl said to him:

> This is sound. You must have a little more respect for the formidable knowledge in technology that you certainly are lacking. Your theory is impeccable. Likewise, your calculations are completely correct. All you need is industrial support to develop such a new combustion apparatus. To which engine builder do you want to go? Perhaps the airframe builder Messerschmitt could be interested. I have family relations with Messerschmitt. BMW? I'll write you any recommendation you want.

Pohl's reputation at Göttingen was sufficient to open the door of any industry for someone he recommended.

Hans von Ohain selected Ernst Heinkel's company at Rostock. Heinkel's company was a strange choice for an inventor with an idea for a new engine because it built airframes but had no engine division. Von Ohain explained his decision: "When I studied in Rostock [in the spring of 1932] I was always seeing the Heinkel Company, which is located there, and got to know a bit about it. Heinkel was the owner of a private company, and had the reputation of being a little bit crazy. I was crazy, too, so maybe we'd be a good team." Von Ohain also knew Heinkel was, as he said, "rich, very rich." Heinkel was a man obsessed with speed—fast cars, fast planes.

Von Ohain gave the rationale for his decision to go to Heinkel. "I was scared to go to an engine company because my model did not work," he said. He felt that the piston-oriented engine industry would scoff at the jet. Von Ohain reasoned, "Those engines with pistons had no kinship with aerodynamics." Von Ohain also liked Heinkel's location near the Baltic Sea. He would try to earn enough money to buy a sailboat.

The family of Wernher von Braun, the rocket expert, lived in Dahlem, close to the von Ohain home. The families were friends. There is no record that von Ohain and von Braun discussed their work with each other during social occasions. It is uncertain if von Ohain had knowledge of the experimental tests of the rocket engine that had been taking place at Heinkel's Marienehe location just before he asked Professor Pohl to write his letter.

It was in March 1936 that Professor Pohl wrote a letter of introduction to Heinkel about his student. Professor Pohl said, "I will write to Heinkel. It will be interesting for me to get acquainted with him."

> First Institute of Physics
> University of Göttingen
> 3 March 1936
>
> Dear Dr. Heinkel:
>
> One of my students, Dr. Hans Joachim Pabst von Ohain, has the unusual ability to make his knowledge of physics technically useful. Some years ago he built an optical interference relay which the firm Siemens and Halske bought at my suggestion, and which the firm immediately placed on the market because of the obvious application for military communications.
>
> Dr. von Ohain has now developed another idea which I consider much more important, namely, a means of propelling aircraft without a propeller and with reduced takeoff and landing distances. He has calculated the process accurately and has begun the practical construction.
>
> I am of the view, however, that the technical development can be carried out only under the auspices of a firm. Therefore I would like to direct your attention to this invention of Dr. von Ohain.

> I assume that you, more than the leader of a physics institute, are pestered by crackpot inventions. Therefore I am writing to you and not Dr. von Ohain so that you may be assured from the outset that Dr. von Ohain's idea is completely serious and firmly based in physics.
>
> With German greetings,
> /s/ Prof. Dr. R. W. Pohl

Heinkel wrote back and said, "Let him come over and explain his ideas." The meeting was set for 18 March 1936.

Von Ohain took the train from Göttingen through Hannover to Hamburg where he changed for the train to Rostock. This train went on to the resort town and spa of Warnemünde, where Heinkel lived. Von Ohain could see glimpses of the Baltic as they traveled the last few miles and thought how happy he would be if he could be employed somewhere near the shore where he could go sailing in his spare time. His dreams of improving aviation did not occupy his every waking hour.

Chapter 4

Meeting with Ernst Heinkel

The importance of Hans von Ohain's decision to take the design of his turbojet engine to Ernst Heinkel is incalculable. Heinkel was extremely successful and experienced as a manufacturer and, most important, had a reputation for taking risks with new ideas. His axiom for success was "Nothing stands still." Heinkel never acted as a mentor or "wise counselor" to von Ohain, but he was an impelling force and financier. An astute businessman, Heinkel set dates when things should be accomplished and was unyielding on his schedules, which focused von Ohain's efforts. With his great wealth, Heinkel provided enormous sums for the development of the first turbojet engines and continued for many iterations that followed. From almost the first day he met Heinkel, von Ohain had resources to build his invention. In contrast, Frank Whittle in Britain took many years to obtain funding to build his turbojet.

Ernst Heinkel was an interesting and unique man, abrasive and persistent in all his dealings. These traits resulted in his accomplishments' being recognized, but won him no friends. In his memoirs, called *Stormy Life* in English translation, Ernst Heinkel says that he was born in 1888, but his life did not begin until 1908, when he discovered air flight. He was a student in the technical school at Stuttgart when he realized that a future in aviation lay in heavier-than-air planes. He says he studied the work of Lilienthal and his gliders and "those liar Americans," the Wright brothers, who claimed they could fly. He read avidly about the new science pursued in France, aeronautics. When he was in his 20s, he built a primitive biplane and flew it without any flight training, with near-fatal results. Undaunted by the wreck, he resolved to design aircraft.

A pioneering pilot who was Heinkel's friend recommended that Heinkel should be hired as chief designer by the Albatros company. Heinkel designed monoplanes, including seaplanes, for military, passenger transport, and cargo purposes. He advanced to chief designer at the Brandenburg Aircraft Works, a company owned by a wealthy Austrian entrepreneur. During World War I he created more than 40 aircraft types, of which 30 were mass produced. Seventy percent of the planes used by the Austro–Hungarian army in 1914–1918 were designed by Heinkel.

Heinkel established his own company, *Heinkel Flugzeugwerke,* in 1922 with headquarters at Warnemünde, and expanded in later years with factories in other locations in Germany. In 1922 the other "aircraft" companies were Albatros in Berlin, Dornier in Friedrichshafen, Junkers in Dessau, and Udet's company in Munich, building gliders and sport planes. Because the market for these aircraft was very limited, these companies were very small. At that time, Willy Messerschmitt was designing and building gliders in a facility more like a workshop than a factory. Heinkel's first designs were civilian and military aircraft for Germany and for customers in Sweden, Japan, and the United States. The innovative designs earned him international acclaim. Some early designs were seaplanes, and Heinkel had the advantage of his location at the Baltic Sea. Seaplanes were popular aircraft because they did not require a long strip of level terrain for takeoffs and landings. In 1927 he became "speed crazy" when he attended the Schneider Cup races in Venice, and by 1932 he won races with his He 64 sports plane. By 1928, Heinkel was producing a great number and variety of types of airframes, including catapulted planes for the German Navy and planes for Russia. In 1928 he first learned of attempts by Fritz Opel to build a car with a rocket engine. In demonstration it had great speed, but it spent more time lifting off the ground than driving on it. It was an idea unsuccessful for a car, but it caught Heinkel's attention. When Heinkel met Wernher von Braun in 1935 with his idea for a rocket-propelled aircraft, he immediately pursued trial flights with the rocket engine.

Heinkel is credited with many inventions. He built an early catapult for aircraft carriers, a device originally requested by Japan for launching planes off shipboard. He invented explosive and countersunk rivets, an improvement that smoothed the airflow over the airframe and thus reduced drag. He is credited with the world's first aircraft ejection seat. He built the first European plane with retractable landing gear. One of his most interesting aircraft designs was a submarine-carried plane, the U 1, that could be dismantled in 22 seconds, fitted into a tank, and reassembled for takeoff in 31 seconds! Examples of this aircraft were purchased by the United States and Japan. In an era of wood-and-cloth aircraft, and inspired by his friend Fokker, Heinkel built one of the first all-metal planes of aluminum alloy and powered it with a Pratt & Whitney engine.

Heinkel's life was dedicated to speed—in his company's aircraft, in his cars, and particularly in the pace of developments in his design office. He was determined that his company would be a leader in the aircraft industry. While German manufacturers were rigidly limited by the Treaty of Versailles as to the types and numbers of aircraft that could be built in Germany as the result of the country's defeat in World War I, Heinkel built what he wanted. Japan was a member of the inspection committee for compliance with the treaty's terms, and would "tip off" Heinkel when inspections were imminent, allowing him to hide his warplanes in the dunes behind the plant for a few days. They were not discovered.

Heinkel had a talent for hiring outstanding employees. He provided strong motivation for them to perform at top level and rewarded them generously from his own fortune. The employment of these outstanding men was fortuitous for future developments at the Heinkel company. Karl Schwärzler, who later provided valuable experience and guidance to von Ohain, had been employed by Heinkel since 1922. In 1931 Heinkel hired Siegfried and Walter Günter to head his design office. They were twin brothers of sensible mien, one a mathematician and one a designer with a flair for designing fast planes. Both were theoreticians who never flew.

These men designed the very successful fast passenger and mail plane, the He 70 *"Blitz,"* as an assignment from Gen. (later Field Marshall) Erhard Milch of the Air Ministry. This plane was to counter the American high-speed planes of the time that were surpassing what any European country had produced. The "Blitz" set eight international speed records in 1933, and it surpassed even its own speed records when it was experimentally fitted with a British Rolls-Royce engine. The British asked for the exchange of some Heinkel airframes for Rolls-Royce engines. The German government hesitated to give another country their best aircraft and restricted the trade to only one airframe. The British used the airframe for tests of many engines.

Heinkel's He 111 all-metal standard horizontal bomber, built at Marienehe, one of the Mecklenburg state parks close to Rostock, and first flown in 1935, flew with telling effect in the Spanish Civil War in the 1930s and participated in the air raid on Guernica in northern Spain, the first big bombing raid ever made against a city. The He 111 variations were a great financial success for Heinkel and were the main German bombers for most of World War II. Some other Heinkel aircraft were the He 177 four-engined, two-propeller, long-range heavy bomber, the *"Greif,"* designed to replace the He 111; the He 219 *"Uhu,"* which was considered the Luftwaffe's best night fighter of the war; and the He 115 twin-motor seaplane.

In 1933 when the *Reichskommissariat* was established, heavy taxes were levied on public companies. Heinkel then assumed full financial responsibility for his company, which gave him total control over the management of the company. During that year, the *Reichsluftfahrtministerium* (RLM) (Air Ministry) wielded its new power and demanded his Warnemünde location as a port for sea planes. Heinkel was ordered to move his factory. The new location was Marienehe. It had 750 acres directly on the Warnow River. Members of the Air Ministry saw, even at that early date, that future wars would be based on air power, and insisted that the factory scatter the workshops apart from each other in order to avoid air raid damage. This campus-like setting was to be the ideal location for later projects when Heinkel wanted to keep the rocket engine and jet engine developments secret.

As Hitler rose to power in 1933, Heinkel had problems in his factories with a few National Socialist employees who insisted that he fly the swastika. A personal call from Göring pointed out to Heinkel that all other aircraft factories were flying the flag, and encouraged him not to refuse. Heinkel, a Nazi party member "for convenience," finally agreed. Through his life, Heinkel was the object of harassment because he was perceived to look Jewish. In the political climate of Germany in the 1930s, he fought off several physical attacks.

Heinkel was a man accustomed to dealing with the upper levels of government. He communicated regularly with members of the newly formed Air Ministry. Self-assured to the point of being arrogant, he had little positive opinion for the technical expertise of those in power at the ministry, and his attitude toward them did not help his cause either in the late 1930s or during the war years.

While Ernst Heinkel was an outstanding German aircraft manufacturer, his place in the world history of aviation was validated by his welcoming and supporting of Hans von Ohain, against incredible odds for success, with his "turbojet device." Erich Warsitz, test pilot, said in an address in 1959, "There has been only one firm in the world which had nothing to build upon, and which found the initiative within itself to go down a new revolutionary path to take that first step over the threshold. That was Ernst Heinkel Aircraft Works, with Dr. Ernst Heinkel himself in the lead."

Hans von Ohain, then 24 years old, took the idea for his "device" to Ernst Heinkel, the head of one of the major aircraft companies in Germany, a man with an international reputation in the aircraft industry. Imagine a 24-year-old with a sheaf of papers under his arm going to the home of the head of the Boeing Company in the United States to tell him, "I have an idea to revolutionize your business." How remarkable that Heinkel listened to him!

Heinkel's home was a sumptuous villa overlooking the beach and harbor of Warnemünde. Heinkel wanted von Ohain to come to his home instead of meeting at the office in order to keep this visit secret from any competitor. Von Ohain spread out his papers with their many calculations and drawings. Heinkel's sharp eyes behind the steel-rimmed glasses fixed on this poised but enthusiastic young man who explained his experimental model. Von Ohain described how the model had been created this far and cited the practical advice and assistance provided by Max Hahn. He urged Heinkel to consider both himself and Hahn to continue in the experimentation with the jet device. Heinkel thought the possibilities were intriguing. He already had the reputation for building the best airframes. He wanted to build the best engines. Heinkel asked von Ohain to identify the problem areas with his proposed engine, wanting to know if this young Ph.D. had a sense of practicality. Von Ohain readily replied that the problems were, first, containing and controlling the combustion, and, second, high fuel consumption. Von Ohain assured him that these could be solved as development efforts proceeded. (Forty years later, von Ohain looked back at his youthful confidence: "I didn't even consider any possibility of failure.") Heinkel asked von Ohain to present the information the next day to Schwärzler, Siegfried and Walter Günter, and his engineering staff at the office in Rostock.

Heinkel saw that the new form of propulsion would require a new design of aircraft. Siegfried Günter reported:

> My brother Walter and I were called to Heinkel's office and introduced to Dr. von Ohain. Ernst Heinkel stated in my presence that my brother would build the airplane, later designated the He 178, to go with von Ohain's proposal. That was a much more important assignment than that for the He 176 rocket plane, whose design likewise came from my brother. Walter had been generally recognized as a good aerodynamicist since he had designed the graceful *Sausewind* (roughly, Whistling Wind)

> aircraft. It was clear that he was the man best suited for the aerodynamic design of novel, superfast aircraft.

The Günter brothers acknowledged the importance of abolishing the propeller for high-speed propulsion. They stressed the requirement that the thrust of the turbojet per square meter of frontal area must amount to substantially more than 2000 horsepower. Karl Schwärzler, the chief designer for Heinkel, said that the turbojet engine would be exceptionally inexpensive to build because of its enormous simplicity, its light weight, and the omission of the propeller.

In 1966, Siegfried Günter wrote a friendly letter to von Ohain, recalling his efforts. It says, in part, "Heinkel understood nothing of the details of your work. [He] understood very well the superiority of a scientific advance, a valuable idea. He often mentioned to my brother and me his extreme interest in your work. On the other hand, in your difficult, risky enterprise you were told to find your own way in all questions."

Von Ohain approached them with confidence, but not with arrogance. He was ignorant of the enormity of the transition from the academic world of theory to the practical world of manufacturing. He recounted the visit in more detail:

> I did not take the model with me; it was too heavy. This was only the first visit at the Heinkel company, and I did have photographs of the model. Heinkel let his engineers look at the photographs and gave me a chance to explain my ideas. This was my first serious contact with engineers. You know, I was a physicist; in a sense, a self-appointed *applied* physicist. The engineers had many thoughts. They told me that the idea of a turbojet was already known, but no one had produced a practical one. On the other hand, they felt that perhaps the time had come to have another try at it.

Von Ohain, in later years, mused about the good fortune of his timing:

> This illustrates the luck of the inventor (or his bad luck). When an inventor is too far ahead of the time he has not a ghost of a chance. But when I came along with my turbojet idea, I consider myself very lucky not to have come too early and not too late. Not too early, because the flying airplanes were des-

> perately looking for more powerful and higher power concentrated propulsion systems for speeds beyond that possible with reciprocating engines. This was not an engineering approach. An engineer would have said, "First prove the principle and forget about weight restrictions." But I tried to make the model light and compact because I had the illusion the thing could propel a plane. I said, "I have made a model but it doesn't work, but I think it has promise."

Von Ohain continued:

> You know, for a full-scale engine with jet flames coming off, you need a special test shed. Heinkel agreed that a test shed could be built. The engineers were not as enthusiastic as I was, but at least they recognized that reciprocating airplane engines and propellers had pretty well reached the limit. At that time the limit appeared to be not over 400 or 500 miles per hour (650–800 km/h), so they thought one should look into gas turbines as a means of raising this limit. They gave Heinkel not a "yes," not a "no." They said, "Well, there may be something to it. It is not as good as he thinks, but it is worth some consideration. It is not a real good machine builder's design, but it is interesting."

Von Ohain also proposed that the engine could be used for the generation of direct lift, that is, vertical thrust. This was possible, he said, because of the high thrust-to-weight ratio with the "pancake" shape of the radial compressor-turbine combination installed in the aircraft wing. (Von Ohain returned to this vertical takeoff and landing concept later, in his years at Wright–Patterson Air Force Base.)

Heinkel quickly made von Ohain a verbal offer of employment on a short-term basis. He wanted to prevent him from going to his arch rival, Messerschmitt. The agreement plunged the young physicist into the high pressure of commercial business, where deadlines are set and expected to be met. He was told he had "a few days" to bring his drawings up to the standards required by the project office. As seen in the following meeting notes, the drawings needed to be completed for a follow-up meeting taking place 14 days later, 1 April 1936.

The following notes from the first meeting were initialed by Siegfried Günter with signed agreement by von Ohain.

Date: 18 Mar 1936
By: S. Günter
Subject: Jet Device of von Ohain
Present: Dr. von Ohain, Schwärzler, S. and W. Günter, Dr. Matthaes, Helmbold
Distr: Dr. Heinkel, Dr. Hertel, Schwärzler, Günter, Dr. Matthaes, von Ohain

1) Drawings and background calculations were presented by Dr. von Ohain.

2) It was found that the thermal efficiency was around 20% and the jet efficiency was about 60% in contrast to an overall efficiency of about 24% for normal engines. Therefore, the engine is most suitable for airplanes that do not have to remain airborne for a long period of time. It was claimed that the thrust can be increased up to around threefold for takeoff by letting the efficiency decrease.

3) At altitude, the overall efficiency remains constant at constant thrust, up to an air density half that of sea level.

4) In the view of Ernst Heinkel Aircraft (EHF), the current diameter of the turbine is very large and would have to be reduced. For comparison, the He 119 with a fuselage diameter of 1 1/2 m (59 in) has a useful power of about 1500 hp, while the turbojet in its current form has about half as much power at this diameter. With some design modifications, however, the diameter could probably be reduced while maintaining the same performance.

5) In the next few days, Dr. von Ohain will modify his design in accordance with aircraft requirements and will present his results in about 14 days. As soon as this material is available, the Project Office will determine the kinds of aircraft for which this apparatus is suited.

Project Office: Follow up 1 April.

6) It was learned that various actual measurements of performance are not available. Therefore, the necessary preliminary investigations would have to be carried out by EHF in any case.

By the time of the follow-up meeting with Matthaes and the technical staff of the research division on 1 April, von Ohain had prepared the drawings, which was enough progress to satisfy the project office. Pressures continued. By the beginning of May "significant progress" was expected. Von Ohain would need to create some convincing data and use some artful persuasion. Four weeks later "positive results" were expected as documented in the following memo. By the end of August, only five months from the start, Heinkel expected any problems with the engine design to have been worked out in order to begin construction of a flightworthy engine! Heinkel was used to setting goals. Furthermore, Heinkel wanted the team to seek advice and assistance by hiring a turbine designer from the steam turbine industry! Von Ohain's creativity might be challenged by a traditional turbine expert. These are notes from the meeting on 2 April.

Date: 2 Apr 1936
By: Dr. Matthaes
Subject: Jet Research
Present: Dr. von Ohain, Dr. Heinkel, Dr. Matthaes, Günter, Dir. Klebe part of the time
Distr: Dr. von Ohain, Dr. Heinkel, Dr. Hertel

STRICTLY CONFIDENTIAL!

1) The most important matter is to determine by tests whether a mixture of hot combustion gas with air can be accomplished in the necessarily short time (about 3/1000 sec) and in the necessarily short distance (around 1/2 to 1 m).

2) These tests are absolutely fundamental to applicability. An appropriate mixer shall be designed and tested to determine the performance and flow patterns in it. The results are the determining factors for applicability: performance and efficiency.

3) Dr. Matthaes is assigned the task of carrying out these tests. Dr. von Ohain will plan to be here from the beginning, preferably right after Easter, and the tests will be conducted to a positive result here.

4) Dr. Matthaes will make another capable engineer, who will be replaced with another man by us, available for the project.

5) If possible, these preliminary tests shall be carried far enough within four weeks that they make possible a decision about the possibility of gas mixing.

6) If the tests have shown a positive result by the end of May, the following program is anticipated:

–Detailed design of the individual components and of the entire machine, especially the blading, above all. Two

> designers and two draftsmen will be needed for this. Of the designers, one must be a turbine designer from the steam turbine industry. An appropriate advertisement will appear in an appropriate technical journal in early May.
>
> –Parallel with this design work, development work must be done on the burner, gasoline delivery, and the blade shapes.
>
> –We must try to accelerate the design and development work with all available means, so that after three months, that is, the end of August, if possible, construction can begin. Under all circumstances, it must be attempted to make the first engine flightworthy.

Von Ohain insisted that Heinkel hire Max Hahn as his assistant. He had worked alone on the updated calculations and explanations, but now he needed Hahn. He appreciated the help Hahn had given him and wanted him to share in the continuing excitement of developing the engine for flight. Heinkel sent this letter.

> Mr. Max Hahn
> Göttingen
> Engelsberg 18
> 2 Apr 1936
>
> I have learned from Dr. von Ohain that you worked closely with him on his jet apparatus. We have concluded agreements with Dr. von Ohain whereby we assume the continued research and finishing of his design. We feel that it is very important that you, having already worked in the area, continue to do so, and it would be very agreeable to us if you could enter into our service.
>
> In case the combustion research that we want to carry out from the middle of April to the end of May produces a negative result, we are nonetheless ready to keep you in our service, subject to agreed terms, because we always need competent skilled workers. It would be very agreeable to us if you and Dr. von Ohain could begin with us right after Easter. Therefore, please fill out the enclosed questionnaire and return it.
>
> Since I am going to be traveling for the next few days, you will want to send the questionnaire to the following address:
>
> Dr. Ernst Heinkel, Feldberg/Schwarzwald, Hotel Hebelhof.

You will receive a response from there immediately.

Heil Hitler!
/s/ E. Heinkel

Max Hahn was given the opportunity to advance from garage mechanic to technician in a major aircraft company on the recommendation of his friend von Ohain and the merit he had demonstrated on the turbojet project so far. Hahn's choice to join von Ohain at the Heinkel Company was a major decision. It showed great confidence in the young scientist and the belief that there would be an eventual positive outcome of the jet engine project. A change of employment such as this was not made lightly in Germany in the mid-1930s because if it did not work out Hahn would be drafted into the army. Assurance from Heinkel that he would always have a job helped Hahn make the decision.

Heinkel chose to believe in the young Hans von Ohain. This was an enormous venture considering that the scientist he was supporting was 24 years old, not an engineer, and inexperienced in the business of manufacturing. There was something special in the poise and confidence and cool reasoning in the young scientist that made Heinkel believe that there was a future to this idea. Heinkel discussed with von Ohain the terms of his continued employment. Heinkel proposed that von Ohain would not be an employee of the company, but work under contract with the resources of the company's top engineers to assist him. He had to strictly promise to make no contacts outside the company concerning the project. He would receive a salary and necessary expenses. The design resulting from the work would belong to the Heinkel Company. Then, if all went well, he would be formally employed with a substantial raise in salary for one year and would be reimbursed for all costs of building the model to date and any costs already incurred in applying for a patent.

This is the agreement signed by Ernst Heinkel and Hans von Ohain:

Dr. von Ohain
Göttingen University
3 April 1936

Dear Dr. von Ohain:

I confirm our agreement of today as follows:

We assume responsibility to carry out the preliminary investigations relative to your patent "Jet Propulsion" according to the attached record, beginning right after Easter and proceeding as quickly as possible, so that by the end of May at the latest it will be clear whether there are any errors or difficulties in your calculations and whether the construction has any prospects for success. Until the conclusion of these preliminary investigations, you will have no dealings with any other organization or person regarding the evaluation of your ideas and device.

During this time, you will be reimbursed for your expenses, such as travel expenses and so forth, and in addition you will receive a monthly compensation of RM 300.—(three hundred Reichsmarks). As long as you are not permanently employed by us, you will be responsible for the taxes on these amounts.

Should the preliminary investigations lead to our continuing the work, we will hire you with a beginning salary of RM 400 gross (four hundred Reichsmarks). The contract is for one year, but at least until 1 June 1937. Should no new agreement be struck by six weeks before the expiration of this contract, it continues implicitly for another year.

In this case, you assume responsibility to entrust your designs and your ideas for jet propulsion only to our firm for reconciling conditions for evaluation, and possibly to convey to us any relevant patents. At our request, you will relinquish to us your claims from the patent filing.

Against that, we assume your complete previous expenses for the construction of the first apparatus and the costs for filing the patent.

We will begin negotiations immediately with your patent attorney about terms for a license with you.

Heil Hitler!
ERNST HEINKEL AIRCRAFT WORKS LTD
/s/ Pabst von Ohain /s/ Heinkel

Von Ohain looked forward to more than the reward of achieving a technical goal. Heinkel told von Ohain, "When you have success, you will be a rich man. I am not selfish with successful people." Von Ohain considered this employment to be a short-term plan. He would provide the engineers his ideas, build a prototype, and return to Göttingen in triumph. "I thought I would stay with Heinkel for a year and then go back to the university," he said. "By that time I hoped to have finished my part of the development, after which the rest would be up to the engineers. It was kind of an arrogant view on the young physicist's part. You have to learn that nobody thinks for you. You have to think it out all the way yourself. I learned it the hard way."

Heinkel assured Professor Pohl that his talented student would be given an opportunity to work on his invention.

Heinkel Aircraft Company
Rostock-Marienehe
2 April 1936

Dear Professor Pohl:

In reference to your letter of 3 March, I have, in response to your urging, already on 18 March and again today spoken with Dr. von Ohain about the matter at hand. I had my best engineers at this meeting, and it is necessary that some preliminary investigations of the combustion problem be carried out. I would definitely like to start on this work immediately after Easter and I hope the work can be finished by the end of May.

I believe it absolutely necessary that Dr. von Ohain be present here during this time, and I would be very grateful to you if Dr. von Ohain could be here from the middle of April to the end of May.

Should our experiments produce positive results and we undertake construction of a unit of that type, it is naturally necessary that Dr. von Ohain remains here at my factory for an undetermined period of time.

I would be very thankful to you if I would get an affirmative answer from you. I am with best greetings

Yours very faithfully
/s/ Ernst Heinkel

Von Ohain's academic success at Göttingen University, along with his friendship with professors, promised that he could get valuable assistance from his former teachers. Professor Pohl was very pleased about the acceptance of his student by the colorful manufacturing genius, Ernst Heinkel. He replied to Heinkel on 6

April 1936:

First Physics Institute
Göttingen University
Prof R. W. Pohl

Dear Dr. Heinkel:

Thank you very much for your lines of 2 April. I fully agree with your position that Dr. von Ohain must be present at your plant during the research, and, should you undertake construction of his apparatus, that he be there constantly. It goes without saying that I will gladly help Dr. von Ohain should he need my advice on some matter.

I would be very thankful if you would report to me from time to time of the progress of the investigations, and it would greatly please me if you achieved a positive outcome.

/s/ R. W. Pohl

Heinkel arranged for von Ohain to work at Rostock in a separate area that was deliberately given a vague name of *Sonderentwicklung* (Special Development), not "jet engine development." Heinkel demanded secrecy from everyone on his staff. On 3 April, Heinkel sent the following to his staff:

Memo for Dr. Hertel

Please see that Dr. von Ohain and his crew get a workshop right after Easter where they can carry out the first experiments according to the attached meeting minutes. In addition, a room for some drawing boards must be arranged exactly like the room next to the Günters, with security lock and so forth.

Furthermore, please see that the people who know anything of this matter sign a secrecy oath. That would be the Günters, Schwärzler, Dr. Matthaes, Bucher, and Gundermann.

/s/ Heinkel

Chapter 5

The hydrogen test engine

Von Ohain settled in to work for Ernst Heinkel at the Marienehe factory near Rostock. Rostock was a charming medieval town on the Warnow River, a center for fishing and shipbuilding. Rostock is considered to be a Baltic port because the river is very wide and deep, navigable by ocean-going vessels. From the docks at riverside, ferry routes went daily to Denmark and Sweden. Baroque four-story houses with stepped gables and Gothic churches surround Rostock's town square. The peaceful area around Rostock had been changed in the late 1920s when Heinkel built aircraft factories. Arado, another aircraft manufacturer, built factories not far away. Von Ohain had attended some classes at Rostock University in the spring of 1932, and he knew there was a strong science faculty nearby if he needed assistance. There were student *Bierstuben* where the young von Ohain could join in the fun and not look out of place.

It was spring, and Rostock was full of activities. There were brass bands on the square, the Whitsun Fair, and tall ships in the harbor. Von Ohain, who enjoyed the outdoors, liked to walk through the town.

The buildings of the Heinkel Company were scattered over an area along the Warnow River. Von Ohain would work in a location far from the main complex, isolating the jet designers from the skeptics in the factory who might express their lack of belief in his idea. A shed with one open side was built for the *Sonderentwicklung* at the end of a long line of buildings rimming a small cove. It overlooked the Warnow where the jet blast could go out over a large expanse of water. A fence about as tall as a man surrounded the shed, barely providing security. There were simple locks on the gates. The area was decreed by Heinkel to be off limits to all but the small staff working with von Ohain. Von Ohain was to report progress on the project to no one but Heinkel.

When von Ohain walked through the charming streets of Rostock and along the shore, his mind was full of modifications and new ideas. He was aware that he faced a formidable schedule to prove to all that his engine would work and his calculations were valid. Heinkel asked von Ohain to estimate the time and cost to develop a test engine. Von Ohain, with the optimism of youth, said that it would be a matter of months and 50,000 marks. That was very much underestimated. It was to take years and a great deal more money.

Von Ohain's education in physics suddenly became inadequate for the task ahead. As a physicist he was well educated to proceed with his turbojet device but lacked knowledge of technical applications found in engineering. Hans stated:

> I knew I had to become a full-fledged engineer. I had to speak their language, to fight with them, and to be recognized as a responsible engineer. This took all my time because when I first came to Heinkel the engineers considered me a crazy boy, a physicist who really didn't know what nuts and bolts were, and who really did not appreciate the problems of materials, materials treatment, castings and honest machine builders' designs. I was indeed completely innocent of understanding in that field.

In a book of cartoons and doggerel poetry made up as a joke by his coworkers in these early months with Heinkel, von Ohain is cited in the *Sonderentwicklungs A B C* with this couplet:

P–"Pabst is ein Erfindermann	Pabst is an inventor
doch er nicht konstruieren kann.	But he can't build anything.

Von Ohain said:

> When I worked with the brothers Günter, I really converted myself completely to what you may call a self-made engineer. I sat down and studied engineering very, very intensively. I knew fluid mechanics very well from Prandtl's lectures. I knew a little bit about stress analysis, so when Prandtl recommended that I get very familiar with stress problems it wasn't difficult for me to apply this knowledge to design of the turbojet. I think that after two years I could cope and compete with any mechanical

> engineer with respect to engineering design. I think I became a very good engineer.

Von Ohain needed to take his idea from prototype to functioning engine. The combination radial outflow compressor and radial inflow turbine was an ideal configuration for a jet engine of very low development risk. In view of the political climate, his tenuous position in the corporation and, most of all, the great impatience of Heinkel, von Ohain could not first develop a well-functioning combustor and then begin an engine design. From previous experience with his first model at the Bartels and Becker garage it was clear that a poorly functioning combustor could result in a nonfunctioning engine. This could mean the end of the turbojet project. He changed to a two-fold approach. In phase one he would quickly build a simple jet engine of minimum risk using gaseous hydrogen as fuel in order to demonstrate the jet principle in a convincing and impressive manner. Von Ohain was convinced that he could win time to systematically complete a combustor for gasoline if the hydrogen engine was successful. Max Hahn could be working on phase two, a fully functioning gasoline combustor.

Von Ohain said that his knowledge of physics brought him to the decision to use hydrogen to run the demonstration engine, but many walks along the beach at Rostock were required before he thought of it. Hydrogen had known properties of high diffusion and flame velocity, plus a wide fuel-air concentration range over which combustion was possible. Von Ohain designed a hydrogen combustor that he was sure would function well and not need time-consuming pretests. It consisted of a large number of hollow vanes with blunt trailing edges placed within the air duct between the compressor stator exit and the turbine stator inlet. Spun parts supported by flanges formed the shroud and flow ducts. Gaseous hydrogen was ducted into the hollow vanes and injected through a number of small holes along the blunt trailing edge into the wake behind the vanes. The risk of this phase was minimized because the hydrogen properties would allow relatively complete combustion before the entrance to the turbine. In this way, the full fuel energy would bc added before any energy was extracted by the turbine. Enough energy was expected to be available to drive the compressor and make the engine self-sustaining.

Von Ohain had freedom and valuable assistance to work out his own ideas. He was the inspiration and the leader. His team consisted of Max Hahn, who worked on manufacturing and combustion experiments; Wilhelm Gundermann, the stress analyst and engineering designer; and two newly recruited draftsmen. Because of the labor shortage caused by Germany's buildup of military equipment, new employees were most often recent graduates from universities or technical schools who lacked practical experience.

Hahn's first reaction to the facilities provided by Heinkel was very negative. Instead of the necessary tools and machinery, there was no foundry, no lathes, no tools to work with the heat-resistant steel for the compressor and turbine wheels. Hahn hastened to identify and recruit the best fitters and mechanics from Heinkel's factory to help him, an action that infuriated the factory superintendents. Hahn played a major part in the design. Also it was his assignment to be working on the combustion chamber. When the combustor blew up, which it frequently did, Hahn worked all night to get it back into working order for testing the next day.

Gundermann's previous experience with Heinkel's company was very helpful. He knew Heinkel's ways and his organization inside out and was able to guide von Ohain through the intricacies and the politics. Von Ohain described Gundermann as an excellent engineer and more: "He was the fixer. He knew the ins and outs of the factory. He could get unbelievable things passed through. If I said to him, 'How do we get this done?' he would say, 'I know so-and-so in that department, leave it to me.'"

Frantic calculations and recalculations began along with recommended design efficiencies. Von Ohain said:

> My greatest attention was devoted to the calculations and layout of the hydrogen combustion engine and to the development of the gasoline combustor. Gundermann and Hahn worked on a design concept using spun parts riveted to ring flanges. Gundermann particularly made the mechanical calculations of the sheet metal rotors and discs. The Heinkel Corporation, the airframe company, was well equipped to produce quickly large spun parts, but was unable to manufacture the ring flanges and the rotor discs. These parts had to be manufactured in a nearby shipyard.

The shipyard was remarkably convenient, located directly across the street from the main factory building of Heinkel's Marienehe plant.

This was Heinkel's private venture. He was an airframe manufacturer and had never before been involved in engine manufacture. This was a bold and expensive venture even for a man who had taken bold risks before. Berlin and the German Air Ministry had total control of all air and related developments through their allocations of materials, but Heinkel said little about this project. Daily, Heinkel would telephone von Ohain asking, "What's new? What's new?" Heinkel knew that Wagner at Junkers Aircraft was working on some kind of jet propulsion for aircraft. Heinkel was also quietly working on another project with young Wernher von Braun, providing an airframe for tests of a rocket engine. Heinkel wanted to be first for any new idea. Von Ohain did not know what to answer. He was fearful that Heinkel would fire him. The political situation at the time was uncertain for young men of his age. The military was building up at an alarming rate. Being involved in a significant manufacturing project could exempt him from the draft.

Winter came. The German winter near the Baltic Sea was bitter, and working conditions were harsh in the little open shed, requiring rugged individuals who believed in their project. The work schedule for the small team was intense. To maintain secrecy, testing was usually accomplished at night or in the very early morning before the factory on the same grounds began work. It is interesting to note that Heinkel's factory, which might have been considered vital to the military, was not, apparently, operating a night shift in late 1936.

The project engine was designated the He S 1, He S for *Heinkel-Strahltriebwerk* (Heinkel jet engine). The He S 1 engine was made essentially of sheet steel fabricated at the Marienehe works and disks created at the nearby shipyard. It consisted of a back-to-back radial compressor and a radial inflow turbine. The rotor diameter was 12 inches (300 mm). The demonstrator combustor did not present any significant problems. The engine operated at a speed of 10,000 rpm and produced a thrust of 250 lbs (1.1 kN). It performed flawlessly under design conditions and during transient acceleration and deceleration. The sound was smooth and focused, and the size of the unit was suitable for an aircraft. Von Ohain tells of the project, "We used a whole battery of hydrogen in bottles and we ran the turbojet from two to five minutes until the hydrogen was exhausted. It was in an order of magnitude of minutes. I had no combustion problems with hydrogen but the metal burnoff was quite a problem."

In a triumph by von Ohain, minor milestones that satisfied Heinkel's concept of sufficient progress were set and achieved, exceeding even the innocent early predictions made by von Ohain. In a new schedule Heinkel stipulated that ground testing of the jet engine must begin no later than June 1937. The He S 1 engine was completed and installed in the test bed about the end of February 1937. Von Ohain said that the first test took place in March 1937. He was firm in his statement:

> Heinkel wrote in his memoirs that the first run of the hydrogen engine He S 1 took place in September 1937. This date is definitely wrong because I remember several comments in which nine months gestation from the beginning to the first run was emphasized. In addition, I recall that water puddles in the vicinity of the jet made the demonstrations very impressive. During March and early April we often had night frost, and prior to our first demonstrations to Heinkel's top engineers and important visitors the test mechanic cracked the thin ice coverage of the puddles.

The following is a description of the test of the He S 1, taken from "50 Years of Jet Flight; Hans von Ohain, an Opposite Reaction," a narrative written by John Hart in the August 1989 publication of *New Zealand WINGS*. Hart's vivid description is taken from an interview with von Ohain.

> The team of three, led by Hans, worked on into the night in the numbingly cold shed. It was wintry inside because a special door on one side stood propped open. Through it could be seen an ice crust glittering on rainwater trapped in the furrows of the ploughed field beyond the shed.
>
> The men checked and re-checked what looked like a prototype agitator washing machine. It lay on its side, bolted to a stand. Unlike the washing machine it resembled, this device had holes in both ends of its "tub." One hole pointed out of the shed's opening.
>
> An old-fashioned leather drive belt, attached to a small electric motor was carefully fitted over a central pulley on the "agitator" which stuck out of the middle of the odd contraption. As the crucial time for testing drew near, one of the men picked up a thick stick and positioned himself behind the leather belt. His

job: whack it off the machinery in the 'washer' at the first sign of the test self-destructing.

The electric motor hummed, spinning the agitator faster and faster. Hans anxiously watched an air pressure gauge connected to the "tub." Its rise told him that the air being sucked into the "tub" was beginning to compress. The "agitator" was, in fact, an air turbine. The shed was filled with a rising whistle.

The crucial difference with the device was that Hans had designed a combustion chamber in the middle of it. A row of small gas burners were fitted inside the "tub" at the point where the air pressure was the highest. Through these Hans was to feed hydrogen gas and ignite it.

Finally, the air pressure gauge in his hand showed that conditions were right in the "tub." Hans opened the gas jet and switched on the ignitor. Unseen, the hydrogen burners began to flare, and the world's first jet roar thundered in the shed. Outside, the ice was lifted from the frozen field and a huge plume of water, mud, and glittering ice fragments billowed off into the night.

The "agitator" at the intake hole in the "tub" was half a turbocharger. The other half was back-to-back inside the exhaust hole. The theory was that the compressed air would be heated by the burning hydrogen gas in the combustion chamber and expand rapidly. Then the heated gases would rush out of the chamber through the exhaust turbine, making it spin. The spinning exhaust turbine would turn the compressor that would continuously suck in air, compress it, and send it through the combustion chamber. The odd-looking device was intended to produce thrust as it hurled out its exhaust gases. Action and reaction, the trick was to get it to run by itself. Without the leather belt.

Suddenly Hans was startled by a shout. The man with the stick was frantically pointing to the belt. It lay on the shed floor where the "washing machine" had thrown it. The prototype jet engine was running on its own power. The von Ohain theory worked!

The young scientist couldn't let himself be distracted. By now the shed was rattling to the jet roar. Hans concentrated like a man with a tiger by the tail. Methodically he fed the engine more hydrogen fuel. It began to speed up. He slightly closed the gas valve and the engine slowed. The test engine was controllable.

> It was the aircraft powerplant of the future that Hans had envisioned as he walked away from his first trip on an airliner.
>
> Hans and his small team had to wait until the engine was almost stopped to hear each other's shouts of jubilation. With wild recklessness they telephoned the Big Boss, Heinkel, to tell him the good news. It was almost 2 a.m. and he was asleep when the phone rang in his mansion. Less than 15 minutes later, Heinkel's massive, steaming Mercedes-Benz sports car slid to a stop outside the tiny shed.
>
> The managing director of Ernst Heinkel A.G., still in silk pajamas (under a great coat), strode in: "Start it again. I want to see this engine of yours running!"
>
> They did. It did. And Heinkel immediately demanded to know when Hans and the design team would have an engine ready for flight testing.

This was a defining moment in jet propulsion, proof that the turbine could extract sufficient energy from the exhaust to power the compressor. Many scientists and engineers had predicted that this was impossible, but von Ohain and Whittle had the right design.

In a lecture later in life, von Ohain gave a description of the sounds of this first engine. "I turned the ignition on and carefully opened the hydrogen valve. The onset of ignition was audible. It sounded like the ignition of a gas water heater." The sound of the world's first turbojet engine designed for aircraft followed. "The engine ran under its own power and accelerated easily. At the higher rotating speeds, it began to produce the sound typical of all later jet engines."

Frank Whittle worked in England with no knowledge of the von Ohain test. In April 1937, one month later, he repeated the German's success with his own turbojet engine, but with kerosene and diesel liquid fuels.

As to the cost of the He S 1, von Ohain said:

> I was interested in the bookkeeping, but Heinkel never gave me a complete picture of it. He said, "This is none of your business." But I could estimate very clearly. I knew all the material costs. I know about how many people were involved in the whole thing. During the first year the relatively low salary of myself was in the neighborhood of 500 marks a

> month, and Hahn's at 400 marks per month. We had a cheap barrack to work in which at today's prices would cost about $2000, without much of a foundation. The parts we needed for the first sheet metal engine I built, I made at a shipyard. They were very cheap. At Heinkel's factory we could deep-draw sheet metal in the forms I would dream up. The forms were made of hard wood and sheet metal was pressed into it. The formed sheet metal was then riveted to the flanges which were machined. So take the salary of two men for roughly one year, and take [the equivalent of] two other men from Heinkel's engineering staff, which was about what we had, and we find that it was a four-man-year effort. A four-man-year effort with a shed, with overhead, in an American factory today amounts to about $150,000 [note: quote is 1975].

The successful run of the He S 1 changed von Ohain from being treated as a clever but probably wrong-headed young amateur to being considered an inventor of genius. His ideas were respected within the organization.

Von Ohain was pleased. "Heinkel and his engineers now believed firmly in the feasibility of the engine, and my position in the company was very firm. It was a considerable morale boost to my coworkers and myself." The simple turbojet fully met performance expectations and also handled very well in acceleration and deceleration. It ran with grace and precision. The psychological effect was enormous.

In March 1937, a patent for the hydrogen-powered turbojet engine was issued.

Von Ohain's one-year contract was changed to permanent employment for *Sonderentwicklung* (Special Development). He no longer worked as part of the research division, but independently with Gundermann as his assistant.

> Director's Office
> Rostock-Marienehe
> 13 May 1937
>
> Dr. von Ohain
>
> 1. We confirm our agreement of today as follows: As of 1 June 1937, you will enter our service and will be in charge of special projects. You are directly supported by Dr. Heinkel.

2. You will receive a monthly salary of RM 500. You must be silent about the amount of this salary, especially in comparison to your fellow workers.

3. We require that you be of Aryan descent and be loyal to the National Socialist state and its leadership. You are willing to confirm this specifically to us in writing. [Note: Von Ohain made the following remarks about this requirement: He asked Heinkel, "Is your company very politically oriented?" Heinkel answered, "This point must be included in every employment contract." In an aside, Heinkel murmured something like, "You will hear no more about it."]

4. This contract is firm until 31 May 1938. From then on, in case we choose not to employ you for further development, we must give you legal notification.

5. You are willing to furnish us a newly issued character reference from the police, which must cover at least the last six months.

6. We expect that your finances are in order, and therefore you do not enter into your service heavily laden with debt. Should you have large obligations, we ask that you make this known to us.

In addition, we cite the general employment requirements and employee regulations which will be furnished to you, as well as the security regulations, all of which are recognized herewith as binding. Should you leave our firm to work for another firm in the aircraft industry, you need a certificate of release from us. To protect your rights, changes to the agreements specified in this document must be in writing.

You are considered a Department Head *(Abteilungsleiter)* in accordance with the enclosed compensation scale. Claims for travel expenses must be based on this scale.

ERNST HEINKEL AIRCRAFT WORKS LTD

Agreed
/s/ v. Ohain /s/ Heinkel
Atch
Compensation Scale

At the same time, von Ohain received agreeable terms concerning licensing and an independent royalty contract. He was reimbursed in the amount of 6000 marks for prior expenses, with 2000 marks paid immediately and another 2000 marks to be paid "when the new apparatus now under construction develops a thrust of 800 kg and weighs no more than 300 kg." The final 2000 marks would be paid "when the first airplane flies with this apparatus."

Von Ohain took advantage of the financial rewards. He bought himself a new car, a Fiat 500 "cabriolet," a two-seater convertible with leather seats. Short trips with his friends to the seaside or the mountains became even more enjoyable. He could be seen on sunny days in his white aviator's cap racing along country roads.

This was found scrawled on the wall at the Heinkel factory:

> Von Ohain fliegt schneller
> Ohne propeller.
> (Von Ohain flies faster without a propeller!)

Chapter 6

Other research in jet propulsion

Air power caught the attention of nations around the world after World War I. The limited amount of flying during the war whetted military appetites for both tactical and strategic uses in future conflicts. Across the world during the 1920s, great improvements were made to aircraft designs that had flown in "the Great War" (World War I), such as lighter materials and lighter aircraft structures, variable pitch propellers, retractable landing gear, cylinder head cowlings, and stronger engines that could achieve 150–250 mph. Experimental all-metal aircraft were built.

However, the Treaty of Versailles that followed Germany's defeat in World War I placed incredible limitations upon the country for building aircraft, particularly onerous in light of their achievements in aircraft design in previous years. The German government signed an agreement in 1922 that no aircraft could be manufactured that exceeded speed, 105 mph; range, 170 miles; ceiling, above 13,000 ft; payload, greater than 1300 lbs; and endurance, greater than $2^{1/2}$ hrs. A few modern "fighter-type" aircraft for the purpose of racing and record attempts were permitted to be built.

Germany became the most air-minded nation in the world in spite of the punishment laid upon it. During the 1920s the country was able to increase its commercial air fleet, including mail and passenger transport planes. A few "military" pilots were trained in Germany, and some were sent to Russia for fighter pilot training. They were the first pilots when the *Luftwaffe* was established in 1933. Sport flying and competitions began to catch on everywhere. Germans established recreational gliding clubs for the sport of soaring. Gliders provided the pleasure and challenge of flight and sharpened the people's thinking in aerodynamics, structural design, and meteorology. Following Lilienthal and the other glider pioneers, Germans made improvements to the aerodynamic structure and controls of the light craft. Their designers showed the world the excellence of German work. Little homemade sports planes began to mingle with the gliders. Stunt flying led to improvements in general aviation.

In the early 1930s, France's air force was the strongest in the world. The pain and suffering of World War I was not forgotten, and the French were particularly bitter because of what they had endured from German invasions in both 1870 and 1914. Germany had a nagging fear that France would seek a war of retribution, but the threat was ineffectual because France was suffering internal political conflict between left- and right-wing factions and was not able to focus on goals outside the country.

In 1933, when the National Socialist Party took command of Germany, Hitler refused to accept any limitation on the development of an air force. Leaders who saw that the future of world power lay in air power devised strategies to restore Germany's dominant position in Europe. With clandestine beginnings, a new German *Luftwaffe* was born, and Heinkel, Dornier, Messerschmitt, and Junkers became its builders. Planes were to be built by the hundreds, not just tens or twenties. German aircraft and pilots of the *Luftwaffe* actively supported the Fascist cause during the Spanish Civil War. The officer corps was instructed to have "a tempestuous spirit of attack." By the end of 1935, Hermann Göring claimed that the German air force had achieved equality with the air forces of Great Britain and France. Hitler announced to the world that he had a military air force. In September 1939, when Germany marched into Poland, the *Luftwaffe* was the most powerful air force in the world. Göring presented statistics on the *Luftwaffe's* size and strength to impress Hitler and to enable Hitler, in turn, to impress the world.

Ernst Heinkel dealt directly with Göring, Erhard Milch, and Ernst Udet, men who were personally involved in policies for the development of the turbojet.

Göring was Hitler's righthand man and air minister. He had been an ace fighter pilot in World War I and had earned Germany's highest military decorations. Described as ruthless and dominating, energetic, sometimes brilliant, and always vain, he was Hitler's principal advisor. Through Göring's influence, a major share of Germany's financial and materiel resources for rearming the country went to the *Luftwaffe.*

Milch was Göring's deputy at the *Reichsluftfahrtministerium* (RLM or German Air Ministry) and dominated in the organizing and building of the *Luftwaffe.* Milch had joined Lufthansa after World War I and in 1932 was a managing director of that company. He was a brash and arrogant man with tremendous drive and knowledge of the production capabilities of the aircraft industry. He understood managers and designers, and had excellent connections with the leadership of the Third Reich. Slow to accept the possibilities promised by the turbojet engine, he liked to give reports of large outputs of equipment to Hitler. Hitler understood reports of numbers of conventional engines.

Udet had the greatest influence on the fortunes of the Heinkel Company. He was the head of the technical office of the Air Ministry, in charge of procurement and development. Göring had selected him for the position to increase the prestige of the *Luftwaffe*, trading on Udet's fame as an acclaimed fighter pilot in World War I and his popularity as a daring competitor in speed trials in the 1920s. Udet was pleasant and friendly but lacked technical and engineering skills and was a poor manager of his office. Udet "fanatically" believed in Heinkel because of the excellent aircraft produced by his company over many years. He provided great help to Heinkel in overcoming problems dealing with the Air Ministry in obtaining materials and supplies for his company, even when he was not convinced that Germany needed the turbojet engine. He was influential in Heinkel's later acquisition of the Hirth Motor Works.

Udet made decisions that had a profound effect on the *Luftwaffe*. He decreed that all military planes, even multi-engine transports, should have dive-bombing capability. In mid-1938, in another decision, Udet's office published an order radically reducing the number of both types and quantities of aircraft and engines. In 1940, Hitler and his generals believed that their war would be short and new developments would not be necessary. The technical office cancelled all government contracts for research and development, stating, "All technical developments which cannot be completed during 1940–1941 must be stopped to release all men and materials for the building of airplane types for immediate use."

Germany's aircraft industry was left without advanced aircraft technology that could have been available to them. The *Luftwaffe* had to complete the war almost entirely with planes developed before 1940.

In 1941 Milch blamed Udet for the ineffectiveness of the air campaign in Britain, and Udet committed suicide.

In spite of the restrictive order published by the Air Ministry, turbojet research continued, with government support for development efforts of von Ohain at Heinkel, and Herbert Wagner and Max Adolf

Müller at Junkers Airframe. The companies were reimbursed for jet engine development costs incurred to date and provided with contracts to continue their efforts.

Hans von Ohain had been appointed Heinkel's director of turbojet engine development. In that position, and as designer of the new form of propulsion, he accompanied Heinkel to meet the technical directors at the Air Ministry, Hans Mauch and his assistant Helmut Schelp. Von Ohain had not only the technical qualifications but also the personal effectiveness to present information requested by the authorities. Through his abilities he earned assistance and encouragement for Heinkel's efforts.

Mauch was assigned in April 1938 by the Air Ministry to head the Special Development Division and investigate "unorthodox propulsion systems" such as rocket propulsion and jet-assisted takeoffs (JATO). When rumors of the mysterious propulsion activities at the Heinkel Company reached his office quite by accident, he made a request to see the rumored propulsion system and to talk to the engineers who were developing it. Mauch, a true visionary, saw the eventual need for an alternative to the piston engine. He quickly grasped both the principle and the importance of jet propulsion and was very impressed with von Ohain. To Mauch, this was the propulsion system for the future. He was very influential in subsequent decisions made by the Air Ministry about jet propulsion.

In August 1938, Mauch met Helmut Schelp, who was in charge of jet propulsion in the Air Ministry's research division. Schelp had been trained in engineering in Germany and the United States and had completed an advanced course in aeronautical engineering at the German Research Institute for Aeronautics (*Deutsche Versuchsanstalt fur Luftfahrtforschung* or DVL) in Berlin, leading to the title of *Flugbaumeister*. This name was awarded to a few men who had shown exceptional ability for future responsibility with the government. Schelp had done independent study of early work on gas turbines, including the French theoretical work of 1910–1920. After considering ramjets and pulsejets, he had concluded that a turbine-driven engine was the best solution for aircraft propulsion. Mauch hired Schelp to join him in the development division of the Air Ministry as head of the *Sondertriebwerke* (Special Propulsion Section).

German research into turbojet propulsion was not isolated at the Heinkel Company in the late 1930s. The Air Ministry was informed of Heinkel's turbojet experiments not later than the end of 1937, although the development proceeded to the fall of 1939 at Heinkel's own financial risk. Herbert Wagner, with Max Adolf Müller, was close after von Ohain with a turbojet design. Herbert Wagner was a brilliant engineer who was well educated in steam turbine design. In 1934, while at the Technical University of Berlin, he conceived the idea for a turboprop for long-range transport aircraft and applied for a patent in 1935. After performing research for the DVL, the official German aerodynamics institute, he was hired in 1936 to head Junkers Airframe Division.

Wagner's assistant in Berlin, Max Adolf Müller, followed him to Junkers Airframe. Müller was a very talented engineer. A review of turbojet patents issued in Germany in 1936–1939 shows as many for Müller as for von Ohain. (In his insistence on secrecy for the project, Heinkel did not encourage von Ohain to obtain patents for his ideas until late in the development of the first engine.) Under secret conditions and with private funds, as Heinkel was doing, Junkers Airframe Division was in the process of developing a demonstrator that had a 14-stage axial compressor, a single combustor, and a two-stage turbine. In 1938, 30 designers and draftsmen were working on the turbojet project at the Junkers machine factory at Magdeburg. Research led by Wagner and his team worked in a logical way, reviewing efforts that had been done before and the patents available. When Herbert Wagner came to ask for government support for the jet propulsion work that he and Max Adolph Müller were carrying out, Mauch was surprised that Junkers was making progress in the same technology as Heinkel.

Von Ohain stated, "By the end of 1938, Helmut Schelp had worked out a detailed plan for the aircraft engine companies to launch a turbojet fighter aircraft development as soon as possible." All types of engines were to be studied or developed in order to determine which one would be best as a service engine. Mauch decreed that all research and development of jet propulsion be carried out by experienced engine manufacturing companies, not by airframe manufacturers such as Heinkel or Junkers Airframe. Mauch considered that the development of jet propulsion would be handicapped by the shortcomings of Heinkel and Junkers Airframe, who did not have experienced engineers and designers. They were building engines in primitive conditions, lacking facilities such as engine and engine component test stands, a materials laboratory, foundries, and suitable machine tools, while the experienced engine companies remained uninvolved. The German Air Ministry was far more interested in the development of jet propulsion at that time than the organization's counterpart in Britain, who had the experimental work of Whittle available to them but chose to ignore it.

Mauch and Schelp saw the future of air power in the turbojet engine. They approached all members of the German aircraft engine industry—BMW, Junkers *Motoren* (Jumo) (this was the Junkers Engine Division), Daimler-Benz, and Bramo (Brandenburg Motor Works)—with funds for the study and development of all types of aircraft jet propulsion.

Some engine companies finally responded to the new idea after Mauch and Schelp went to them with generous offerings of money and facilities. Contracts were reluctantly accepted by Jumo and Bramo-BMW. Daimler-Benz flatly refused. The government had identified that Germany's piston engines were inferior to those in Britain and the United States and had overloaded the companies with contracts for improving them. The manufacturers hesitated to enter the new area of research because rapid build up of the munitions industries throughout Germany had caused an extreme shortage of engineering and technical assistance, as well as materials. Their reluctance to accept the new technology is understandable considering the practical difficulties that might have come with unsolved problems and expensive retooling.

Under the leadership of Wagner at Junkers Airframe Division, the Wagner-Müller turbine engine continued in development with private funding. The Junkers Engine Division continued with improvements to the reciprocating engine using a propeller. In late 1936, Junkers Airframe and Junkers Engine Divisions became one entity. Professor Otto Mader was assigned as head of engine development for the consolidation. He caused a long delay for any further work on the turbojet project because he stated strongly that "the nonsense of jet engines" would never come into his company. In 1939, the Air Ministry pointed out to Mader that he already employed Anselm Franz, who would be an excellent team leader, and Mader conceded to let him proceed. Franz's first action as leader was to declare that the Wagner-Müller engine was not worth further development. He had previously worked on jet concepts, but not a turbojet, and now had an idea for his own design. His work resulted in the Junkers Jumo -004 engine that was later used in Woldemar Voigt's design of the Messerschmitt Me 262 aircraft, the world's first operational jet fighter.

At Bramo, a team led by Herman Oestrich began work on an axial flow turbojet engine with the support of the Air Ministry's technical office. In 1938 Bramo and BMW worked together in development, and in 1939 the two companies were merged. With the Bramo-BMW merger, this engine became the BMW -003 engine, intended for use in the Messerschmitt airframe. The engine was expected to be available for installation in the airframe by December 1939. It was later used in the Arado 234C and the Heinkel He 162 aircraft.

Mauch wanted von Ohain to take his team to Daimler-Benz and offered Heinkel reimbursement for his investment in jet propulsion made to that point. Daimler-Benz agreed to the transfer, but Heinkel was adamant that the effort be continued at his own company. Von Ohain agreed. He wanted to continue with Heinkel because with Heinkel he had complete technical freedom, no financial pressures, no government requirements, and no time delays. There were aircraft waiting for his engines.

Karl Leist at Daimler-Benz was working slowly on a turbine-driven ducted fan for an engine of advanced design.

There was one other turbojet engine that could be developed. The team that had worked on the Wagner-Müller engine at Junkers Airframe believed in the engine and resented its abandonment in 1936 at the time that the Aircraft and Engine Divisions merged and Franz was placed in charge of jet engine development. When members of their team were being reassigned, Müller, the highly regarded compressor specialist R. Friedrich, and others who had worked together advertised themselves as a team available for engine work. Heinkel made them attractive offers in 1939 and they joined the Heinkel Company, bringing with them the axial engine (Heinkel-designated He S 30) along with a gasoline engine and a diesel engine for further development. Heinkel was pleased to gain a team working on an axial engine.

The Air Ministry formally gave each turbojet development a government identification number and stated the scientist in charge.

The 1939 Air Ministry Program

RLM Designation	Contractor's Designation	Chief of Development
109–001	Heinkel He S 8	Dr. v. Ohain
109–002	BMW P 3304	Weinrich
109–003	BMW 003	Dr. Oestrich
109–004	Jumo 004	Dr. Franz
109–006	Heinkel He S 30	M.A. Müller
109–007	Daimler-Benz ZTL	Prof. Leist

Even before the first von Ohain jet flew, Heinkel found himself with two strong competitors who had initiated jet design projects, Anselm Franz at Junkers Motoren (Jumo) and Herman Oestrich at

Bramo-BMW. Each had enormous competence in aircraft engine design, and each was backed with experienced engineers and excellent facilities. Both companies started out with much stronger teams than Heinkel's and grew rapidly in strength. They also had full support of the Air Ministry in every respect including facilities and equipment. The Heinkel Company persisted with jet engine development. In spite of his limitations Heinkel had one great asset, Hans von Ohain. He cried out to the young inventor constantly to get the engine ready to fly, stressing, "We must be first, we must be first!" Heinkel's concern was not unfounded.

Names in other countries can be cited for their turbojet research. For example, the Italian Caproni-Campini CC-2 flew in August 1940, although it was not truly a turbojet. The Soviets had their own turbojet heroes including A.M. Lyu'lka, who began work on aircraft gas turbine engines in 1937 and completed the VDR-3 turbojet in 1945. In the 1930s, Sweden and Japan initiated work on turbojets. The United States began their efforts in 1941.

Chapter 7

Heinkel's engine developments

Both the rocket engine and the turbojet engine work on the "jet" principle. They create a thrust forward by accelerating a flow of gases through a rear nozzle.

A rocket engine performs independently of atmospheric air because it carries its own propellant containing both fuel and an oxidizer. The duration of its performance is limited by the mass of propellant that can be carried. It is not uncommon for a rocket-powered vehicle to be 90 percent propellant by weight.

A turbojet engine is based on the gas turbine principle. It is dependent upon the atmosphere. It works on principles of induction of air, compression, combustion, expansion, and exhaust. Its inlet scoops in enormous quantities of air delivered to the compressor, where it is compressed as much as 20 times over the pressure at the inlet. The compressed air passes to the combustion chamber where fuel is injected and burned, creating a mixture of hot air and combustion gases. A large part of the energy of these hot products is then extracted by the turbine, which turns the compressor and all the necessary accessories. The remaining energy of the exhaust gases is used to create a strong jet, exhausted through the nozzle at the rear of the engine to create thrust. The thrust is equal to the mass flow rate times the difference between the exhaust jet velocity and the flight speed. A turbojet engine can be throttled over a wide range to control thrust, which is not the case with the rocket engine, which is more limited.

Ernst Heinkel was unsurpassed in the quality of airframes he was supplying to the *Luftwaffe* and unrelenting in the quest for ways to improve the speed of his aircraft.

In early 1935, even before he had heard about the young student, Hans von Ohain, Heinkel saw a demonstration of a rocket engine designed by an unknown young scientist named Wernher von Braun. Heinkel remembered the Opel rocket-powered automobiles from 1928 and immediately thought that this engine could be developed for aircraft propulsion. The German Army was not interested in testing it in an aircraft, and the Air Ministry saw no use for it as a new form of propulsion. When von Braun asked Heinkel if he would supply an airframe for the flight demonstration of the engine, Heinkel quickly agreed to provide his He 112 airframe. Because of frequent engine explosions, Heinkel had to provide several airframes. Von Braun was never employed at Heinkel but stayed with the promising German rocket program that made history of its own.

In 1937, Erich Warsitz, a test pilot, was hired to continue development of a rocket engine in a secret testing area at the Heinkel Company. Warsitz had been a motorcycle race driver before he began to fly. He became an instructor at the School for Civil Aviation in Stettin, secretly teaching multiengine pilots, before transferring to the clandestinely forming *Luftwaffe* to work with von Braun on development of the rocket engine. The glamorous Warsitz was tall, blond, and fearless. Von Ohain described his appearance as "the Hollywood model of what a test pilot should look like." Mortality was high for test pilots in those days. Warsitz was a bachelor and there was less concern about hiring him for the job. He explained, "There was a lot of work for test pilots. The only way we got any time off was to be in an accident and land in the hospital."

Heinkel's team designed the He 176 airframe especially for testing of a rocket engine. It had been made as small as possible to be able to shatter all existing speed records. The current speed record was about 460 mph (750 km/h), and the He 176 was designed for 600 mph (960 km/h). The wingspan and fuselage lengths were equal, approximately 17 ft. The plane was 5 ft high and the fuselage 2 ft in diameter at its widest point. Warsitz found he was slightly cramped if he forgot to take out his wallet before he got in. The pilot was required to recline in the plexiglass-enclosed nose. Behind him were the fuel and the engine. Heinkel's engineers had designed a jettisonable nose section that could be ejected if the pilot had to leave the plane. It would be slowed by a drogue parachute, allowing the pilot to use his personal parachute. The power for the He 176 was not provided by the von Braun engine but by the less powerful and less explosive HWK-R I 203 rocket engine designed by Helmuth Walter. Fuel limited powered flight to only 30 seconds.

On 20 June 1939, Warsitz had his first triumph for the world record books as pilot of the first successful rocket-engine aircraft. The flight was a startling 50-second vertical loop, a somersault from takeoff to landing. On the following day it was demonstrated before Generals Udet and Milch. The Air Ministry representatives thought the whole thing was a stunt. It was, however, put on the secret list. Udet professed horror at the thought that Warsitz was flying in such a dangerous craft and banned further flights. Milch promoted Warsitz to flight captain on the spot and gave him a pay raise at Heinkel's expense. Udet later relented, under persistent pressure to allow further flights.

An order was sent by the Air Ministry to all aircraft manufacturers. On 3 July 1939 all advanced airplanes were to be shown to Hitler, Göring, and other officials. Heinkel was asked to take the He 176 rocket plane to the Rechlin Proving Ground for a test flight. The He 178 airframe for von Ohain's engine was taken for static display. Udet asked for a demonstration of the rocket plane. He told Heinkel, "The Führer's got to see something new, so I remembered your funny kite. If it can get around the airfield, that will be quite good enough."

In his autobiography, Heinkel said:

> Hitler was accompanied to the airfield by Keitel, Jodl, Göring, Milch, Jeschonnek, and Udet. He came over and gravely shook hands with Udet and myself. He examined the plane for a moment, listened to Udet's hasty explanations, and then stepped back about a hundred yards with his staff. I held my breath as Warsitz squeezed himself into his seat. A stream of smoke shot from the tail of the plane, and the birdlike craft moved off across the airfield. When the plane was airborne, the undercarriage retracted. At a height of between two and three thousand feet, Warsitz circled above us, throttled back, glided down to the ground, then suddenly switched on again, so that the plane shot once more into the air. At last, he closed his throttle and landed at a speed of nearly 200 mph. The plane taxied across the airfield.

Heinkel had hoped to impress the group with his bold new form of propulsion, but there was no response. That same day Hitler and members of the Air Ministry staff were shown the He 178 airframe in static display. They were told that there would be a flight propelled by a new form of propulsion, a turbojet engine, within a few weeks. Hitler asked, "Why do you think you need a new engine type to fly faster than the speed of sound? Why can't you do that with a normal reciprocating engine?" Rather than a full explanation, they gave him the most simple. "The propeller loses efficiency at supersonic speeds and consequently the engine gets too heavy. A new lighter engine development is needed." Hitler made a few complimentary remarks to Heinkel and took his leave. There was no particular interest shown. Von Ohain heard later from Hitler's adjutant about Hitler's reaction. "The demonstration took place too early in the morning for him, hence his cold disposition. He preferred not to get up before 11:00 a.m."

Göring asked Warsitz his opinion of this new development. Warsitz said, "I am convinced that in a year or two very few military planes will have propellers and reciprocating engines." Ignoring the possibilities of what the new propulsion could offer the *Luftwaffe*, Göring smiled good naturedly and said, "You're an optimist."

By now Ernst Heinkel thought that the Air Ministry's lack of response to his innovations was a plot by the powerful group of manufacturers of traditional reciprocating engines. He believed they were against any new engine developments that were not products of their own efforts. Heinkel was convinced that his turbojet engine would be an outstanding success. Although it used enormous quantities of fuel, it was lighter and easier to produce than the reciprocating engine and it answered the need for higher speeds. Unlike the rocket engine, its thrust could be controlled. Heinkel reasoned that if this new engine in its new airframe could earn the support of the Air Ministry, they might permit him to invest in an engine factory. They might even be forced to give him a factory to build engines. If he could stage a demonstration of the first jet plane and jet engine before an audience of Hitler, Göring, and such members of the German Air Ministry and the *Luftwaffe* who might be assembled, they would applaud his genius. It would show his foresight, the brilliance of his judgments of new ideas, and the remarkable capabilities of his staff. He would outsmart the German engine companies of which he was so contemptuous.

Heinkel provided a continuing contract to von Ohain. The salary was lower than the amount he would pay to another engineer in charge of such a development because, he said, "I don't know you yet." He promised that when von Ohain's engine propelled an aircraft there would be an increase. Von Ohain was satisfied with this arrangement and considered it to be "fair and square."

Heinkel did not understand the difficulties of developing a turbojet engine. In his youth he had built wooden aircraft with fabric wings, rapidly constructed and rapidly rebuilt, and believed that building an engine required the same kind of effort. Heinkel liked to lecture that success was the demonstration of the ability to build something, and after which you could ask for more support and achieve further objectives. Von Ohain had, in a way, met that criterion by having a model engine to show Heinkel when he first asked for support (although it didn't work). The turbojet engine was a simple concept, but not simple to build. It needed engineers and specialists in materials, manufacturing and casting processes, and controls. Heinkel was using his best personnel to continue work on the airframes he was producing in quantity and could not find others to hire because they were being used in other factories for production of conventional weapons for German war materiel. Von Ohain had to be satisfied that his components were built and accessories selected by relatively inexperienced engineers.

The responsibility for the success of the entire jet engine program lay with von Ohain. Von Ohain told the story: "I jokingly said to Heinkel, 'You want us to build a precision telescope out of beer bottles, and when we are successful we will get the necessary materials and grinding machines.' Heinkel laughed and nodded, but said nothing. He had a good sense of humor and was not angry with me."

Von Ohain laid out his plan:

> The gasoline combustor program:
>
> 1) Installation of a two horsepower Sirocco blower with controllable bypass.
>
> 2) Investigations on segments of annular combustors
>
> –Influence of the shape of the combustion chamber
>
> –Flame holding mechanisms
>
> –Factors influencing combustor volume for given pressure and fuel flow
>
> –How to obtain a low total pressure drop through the combustor
>
> 3) Gasoline vapor generation and injection into the combustor, that is, generation of high pressure gasoline vapor by an electrically heated pressure boiler.
>
> 4) Combustor utilizing atomized liquid fuel.

Some problems were unique to the jet engine. Von Ohain's bent sheet-metal parts were inadequate to withstand the extreme heat of the burning mixture of compressed air and fuel. Materials to withstand the hellish temperatures seemed to be nonexistent or at least unavailable, and nickel was in particularly short supply. They had to devise a way to feed the fuel. The team did not have fuel pumps that could handle the necessary quantities of fuel, but they thought they could substitute pressurized tanks for a first flight. Some experienced help would have solved the easier problems. Von Ohain said:

> We were not, for instance, familiar with bearing lubrication or its problems or, in general, with good machine design techniques. We had not yet solved the problem of how to inject the fuel or how to make gasoline burn properly in a combustor. I began to realize that the turbojet was not just a flying can with sheet metal wings, but a machine, after all. Many components were not compatible with each other because we did not have the results of component tests for compressors or for turbines. Eventually we were able to calculate our component efficiencies to match the design point. Later on when we talked with people in DVL they said that this could not be done without test facilities and component testing.

Von Ohain laughed with great pride but little humor. "But we did it."

Heinkel provided funds for von Ohain to seek advice outside the Heinkel Company and told him that if he did not reveal the scope of his project, he could go anywhere he wanted for help. Von Ohain sometimes disappeared from the factory for a few days and visited one of his contacts with some problem that had to be worked out. Von Ohain, the scientist, continued to confer with his connections at Göttingen University, with the research facilities in Göttingen, and with former school friends. He was able to visit reciprocating engine experts. He became acquainted with engineers in the combustion equipment industries and went to industrial exhibits. He visited Lilienthal Society meetings where experts on soaring could provide their ideas on airframe design. He visited the Leipzig Trade Fair to find a manufacturer who might have materials that could be used, but with no success. Von Ohain was respected for his knowledge and his honesty, and what assistance was available was unhesitatingly provided.

There was a need for unconventional solutions for the problems he identified. In previous scientists' experimentation with turbojets, combustors had been the main barrier to successful development. The size of a unit using conventional combustion technology would be as large as a good-sized room. Even today when jet engines have achieved great sophistication, the design of combustors is a primary concern of engine companies. For von Ohain and his team, the problem of an adequate combustor had been almost abandoned in order to get the He S 1 hydrogen engine demonstrated for Heinkel. Developing a combustor was now taking most of the time of von Ohain and Max Hahn. Any scheme was tried. As von Ohain said, "We used a pressure cooker of the kind used to boil potatoes. We boiled gasoline and used the steam for the combustor. That's probably not consistent with today's standards of safety, but we had our own safety ideas. It was a little crazy."

Gasification baffled the team until the pragmatic man, Hahn, used his experience. Von Ohain tells the story:

> Hahn had no training in physics or chemistry as I did, but we worked together to come up with something which was practical. He said to me, "This fool science doesn't help you. See here. I have a soldering iron, a soldering torch [a blowtorch]. You see it gasifies its fumes. Let me make a combustor like this." We built what could be called a gigantic soldering torch by making tubes for the vaporization of the gasoline. A fan was used to stimulate vaporization. This was followed by the combustor where the gasoline vapor was burned. Pretty soon we had a little combustor outside the engine. We got beautiful combustion. The only trouble was that I was afraid of what might happen with the accumulations of carbon deposits on the vaporizer tubes.

Von Ohain explained further:

> Hahn suggested that we approach the liquid fuel combustion development in two steps: first, to use vaporized fuel; later, atomized liquid fuel. Hahn built an electrically heated pressure boiler with the necessary safety measures for producing high-pressure gasoline vapor. After achieving combustors with excellent operational characteristics and low total pressure drop, the boiler elements were built into the combustor. We developed, after much experimentation, a vaporizer of gasoline fuel built like a boiler, and then put the vapor into combustors which were similar to propane combustors. We were able to operate

> the engine for about five or 10 minutes before carbon deposits interfered with the vaporization of the gasoline. It was all very fantastic when I think back on it.

Max Hahn suggested the idea of arranging the combustor in the large unused space in front of the radial compressor, pointing out that this would greatly reduce rotor length and, in addition, total weight. Von Ohain said, "I could see many additional mechanical and aerodynamical advantages. I incorporated Hahn's suggestion in the layout of the flight engine and worked out the aerodynamics of the air ducts and the mixing of the flame gases with the bypass air." An international patent for the combustor was awarded solely in the name of Max Hahn. Von Ohain was elated to see his good friend achieve this recognition. It was a generous act on the part of von Ohain, who had made a considerable contribution to its design.

Ernst Heinkel as sole owner of his company was able to absorb the political waves pounding the aircraft industry and keep them from interfering with the work at his factory. Heinkel approached Udet in December 1938 to ask permission to seek foreign patents for the jet engine, but he continued in his insistence that the extent of progress on the turbojet engine should be considered secret. The airframe workers at Heinkel's factory were not told the reason for the roaring sounds and flames that appeared in his Special Development Division and became the subjects of factory jokes. Whenever top Nazis or members of the Air Ministry visited the plant, Heinkel carefully steered them away from the secluded areas. Even the Party members he was forced to hire in his plant were kept unaware of the special engine project. By keeping the work secret, Heinkel missed the opportunity to get experts from the Air Ministry and experienced engine companies to help as consultants.

Secrecy of progress was uppermost in Heinkel's plans. This is his memo dated 5 January 1939:

> Explanation of Obligation to Maintain Secrecy of the Special Work of von Ohain with Regard to Visitors From Outside the Company
>
> In order to protect the development work that we have achieved with the von Ohain device, I place upon each of you

> the duty to maintain absolute silence about the status of our development, whether in the plant or at the Ministry, and also to show the designs or the completed machinery to no visitor who does not have my personal authorization.
>
> I alone will decide who has the right to visit the Special Development. In every case, therefore, my permission must exist. Only on the authority of a current authorization may Dr. von Ohain allow entrance to an outside visitor.
>
> If I am on a trip, no viewing of the design or the hardware can take place.
>
> /s/ Heinkel
>
> Read and Understood:
> /s/ Names of 9 scientists.

Heinkel enforced the earlier directive with the following memo to the same individuals on 13 February 1939:

> In many places, Special Development II (von Ohain) is viewed only as a company secret. According to a written declaration of the RLM, it is indeed also a State Secret. Therefore, the provisions of Paragraphs 88-92, 139, and 153 of the Book of Federal Penalty Law apply to everyone whose work is related to Special Development. Moreover, I point out that no one has the authority to tell any other person about the work of Special Development without my express permission.
>
> /s/ Heinkel

Ernst Heinkel appreciated the talent his young protégé exhibited in spite of Heinkel's constant nagging for a date when the engine would fly. It was 29 June 1939 when Ernst Heinkel published the following memo:

> Memo No. 141/39 from Dr. Heinkel
>
> Subject: Establishment of a Museum for Dr. von Ohain's Jet Engines
>
> At the beginning of next year, we will probably have a building free where we can fashion the beginnings of a museum. In this museum, I want the total development of the jet engines from the beginning of 1936 with Dr. von Ohain.

> All jet apparatuses from the beginning are to be preserved, even the half-destroyed.
>
> One example of an engine is to be cut away so that the inner construction can be seen.
>
> I want to be informed as soon as possible that the hardware still exists, and also how many there are, what models, etc, and also find a room where these engines can be saved as of now.
>
> Responsible for carrying this out:
>
> For assembling the hardware v. Ohain
> For finding the room Kapp
> For the display and protection Kettel

Von Ohain was not one who talked about his work, even to his family. One weekend when he was visiting at home, his parents wanted to talk with him about this big project that seemed to be taking all his attention. The family customarily took long walks in the forest on Sunday mornings. In this relaxed atmosphere they believed that their son would tell them about what he was doing at the Heinkel Company. They broached the subject while strolling, and he reluctantly replied, "I'm working to build an engine that doesn't make any sound." The family was more puzzled than before. The statement can be considered accurate. Conceptually, all you hear in jet flight is the humming of the turbine. Or to borrow from the vivid description of the first flight of the Army Air Corps/Bell/General Electric XP-59A at Muroc Dry Lake, California, on 2 October 1942: "What a strange feeling....dead silence as it passed directly overhead...then a low rumbling like a blowtorch...and it was gone...."

The He 178 airframe was intended as an experimental vehicle to demonstrate the principles and characteristics of jet propulsion, not to be the prototype for a production aircraft. Provisions were made and space left available, however, for the installation of gun armament.

Design of the airframe was initiated in the summer of 1937 by a team at Heinkel led by Heinrich Hertel and chief designer Karl Schwärzler, together with the Günter twins and Heinrich Helmbold. Hertel had given the Günters a completely free hand in aircraft design at the start. Detailed sketches of the design were on the desk of Walter Günter when he was killed in an automobile accident in September 1937. Helmbold, his closest coworker, carried the design to completion along with Walter's twin brother, Siegfried.

The aircraft had a duralumin monocoque fuselage with a bubble-type canopy over the pilot. The wings were set high in the fuselage in an elliptical wing planform, a Heinkel trademark. Hertel later forced the use of a trapezoidal wing, speaking as a stress specialist, not as an aerodynamicist. The size and shape of the air intake to the engine, the position of the engine, and the shape of the jet outlet at the back of the plane were entirely new considerations for an entirely new aircraft. For example, because there was no propeller there would be no torque, and no trim tabs were provided. Gundermann contributed to the design of the air inlet and the air and exhaust gas ducts. The engine assembly, with an intake "mouth" shaped like a striking trout, ran from the nose, under the pilot's seat, to the exhaust nozzle at the extreme rear of the fuselage beneath the rudder. The tailpipe was fairly long, about one-third of the aircraft's total length. A narrow landing gear was designed to retract into the fuselage. Initially, the plane was to have two engines, but Heinkel opposed the idea as too expensive. The engine would be placed behind the pilot with its forward end roughly even with the wing trailing edge. A single fuel tank was immediately behind the cockpit.

Concurrent with construction of the He 178 aircraft, von Ohain completed the design and was constructing what he believed would be his flight engine, the He S 3A. In early 1939, Reidenbach and Antz of the Air Ministry inspected the von Ohain engine and the mockup of the He 178. The engine's performance was still too low, producing only about 800 lbs of thrust when a thrust of 1000 to 1100 lbs was estimated to be necessary in order to take off from Heinkel's relatively small company airfield. The team made several improvements, mostly refining the easily exchangable radial cascades of compressor-diffuser and turbine-stator. Von Ohain worked on a design for an experimental engine with an axial compressor as an alternative engine for the He 178.

By March the prototype engine achieved 992 lbs of static thrust. The engine was bench tested and later flight tested under a Heinkel He 118, a prototype single-engine dive bomber. In his autobiography, *Stormy Life*, Heinkel stated:

> The machine stood high enough to give sufficient ground clearance so that the jet engine could be slung under the fuselage. Warsitz flew with Kunzel, an engineer. After the He 118 took off with its normal engine, Kunzel would switch in the jet unit. It was a brief but indescribable performance that we saw. We used to watch how, first, the bluish jet stream shot into the

> air, and, almost at once, the plane was forced up to an enormous speed. These air tests were repeated day by day until the early refractory metal of the turbine burned out one day, fortunately as the aircraft was on a landing approach.

The fire also burned the He 118 airframe. Work continued on a duplicate engine, the He S 3B.

The design of both the He 178 airframe and engine had been set by the end of 1938. Difficulties with the prototype airframe were not the responsibility of von Ohain, but the engine required long hours of work for him and his team. Von Ohain said:

> A tremendous pressure was put on me. Heinkel said it was not important to have a real mass production engine. The important thing was to have a successful flying engine even if it was a temporary success. Regardless of the quality of engineering, Heinkel was determined to be a turbojet manufacturer. "When we have this," Heinkel said, "it will be easy to get proper support (from the Air Ministry)."

Heinkel was determined that the government should help him acquire a factory to build turbojet engines.

All problems were not yet solved. Von Ohain said:

> The final difficulty of the flight engine lay in the need for achieving high mass flow and high component efficiencies of compressor and turbine. The high mass flow was obtained by an unconventionally large ratio of compressor rotor inlet diameter to rotor exit diameter. Normally such a compressor configuration would result in very large inlet losses caused by too high Mach numbers and too large inlet blade curvatures. I tried to reduce these losses by means of an axial inducer stage which gave the inlet flow both a precompression and a prerotation, thereby substantially reducing the Mach number and curvature of the rotor inlet blading.

Some of the air from the compressor passed forward into the reverse flow annular combustion chamber while some went rearward to mix with the combustion gases before entering the turbine.

The 27-year-old von Ohain alternated between elation and discouragement. Heinkel's daily inquiry about how work was progressing spurred him on while daily problems filled his mind. The reality of his goal of turbojet flight seemed to be within his grasp, with smoothness and efficiency in production of power, and "elegance." He thought again of the possibility of taking this triumph of creativity back to Göttingen University as criterion for being hired as professor, yet he saw that he could make continuing contributions to the science of turbojet propulsion.

Much was learned from the many test flights of an engine attached under the He 118. By early August 1939, after additional modifications, performance of the He S 3B engine very nearly reached the required thrust. A bench test of 10 hours qualified it for a flight of one hour. It would be installed in the now-completed He 178, the world's first pure jet, in preparation for taxi tests. Erich Warsitz was selected to be the pilot for the first flight. In spite of Heinkel's quest for complete secrecy, Warsitz was permitted to make a few taxiing tests and soon declared that he was willing to fly. Although problems and adjustments to the engine were not all resolved, the decision was made to fly "the next day."

A ground crew was sent out to the full length of the grass runway to stomp down any uneven heaps of ground and fill any depressions, in order to have the smoothest possible path for the takeoff. Ingestion of a bird into the jet engine was a problem for even this first flight. During taxiing tests, a bird was sucked into the engine and the engine had to be taken apart. Minute, well-cooked traces of the bird could be found in the undamaged turbine. "This took an hour or two," von Ohain chuckled as he told about it 40 years later. "Of course, that engine was much less complicated than jet engines of today."

Chapter 8

First flight of a turbojet-propelled aircraft

Sunday, 27 August 1939 was a beautiful morning by the Baltic Sea. Von Ohain recalled, "The sky was dark blue like an Italian post card. There was a strong early morning fog, but soon the fog started to move. It was one of Mother Nature's most beautiful days." At that latitude the first light was about 0500 hrs, and the team of von Ohain, Max Hahn, Karl Schwärzler, and Siegfried Günter wore light jackets in the cool air. Riggers from the experimental section were already in place at the building where the He 178 with its He S 3B engine was being readied for flight. Test pilot Erich Warsitz paced nervously. He remembered, "I had given instruction to one of the engineers to forward a letter out of my desk if anything fatal happened. I had said a special goodbye to my dog, as I did each time I had dangerous tests." The team waited for the car that would bring Ernst Heinkel.

In the cold light of dawn von Ohain paused to think of his dream of elegance in flight. Much work had been accomplished and many things had gone right. Now, at the rising of the sun, his engine might lift the fragile airframe Heinkel had ordered, and flight would never again be quite the same.

Heinkel arrived. The plane was towed from its shed to the takeoff point by an automobile longer and bulkier than the plane. The cold engine was primed with hydrogen to get it started. Warsitz said he "tested that the ailerons were working, checked the turbine at different revs, checked the pump pressures, temperatures and lots of other things." Heinkel rushed to the plane and grasped the gloved hand of Warsitz to wish him success. The mechanics got the signal to close the cockpit. The now-familiar high-pitched whine from the engine's turbomachinery echoed across the field.

In an article in the German publication *Luftfahrtgeschichte* in September 1991, Adolf Jensen, who was responsible for the aerodynamics of the He 178 aircraft, described some events of the day.

Heinkel sent Jensen to his car, which had a radio installed. The news was on, and the air was filled with rumors of war. When Jensen reported what he heard, Heinkel said, "I hope those idiots don't start a war, so that we can proceed with our work." Jensen said he remembered very well what Heinkel said because it was risky to make such a statement in those days.

Jensen wrote about a design problem unique to the jet aircraft. In a propeller-driven aircraft, the propeller airflow ("prop wash") over the airframe flows to the rudder, providing control for rolling maneuvers. In the case of the jet aircraft, there is no "prop wash" over the rudder, and side to side control from the rudder has to be connected to the wheel brakes in order to enable the aircraft to hold its starting direction. Warsitz's rolling tests showed that the brakes needed adjustments, which were quickly made.

Because wind-tunnel data did not exist, uncertainties remained, particularly the adjustment on the stabilizers. The aircraft was poised at the end of the runway for takeoff. After a run of 300 m, speed built up. Hot exhaust gases scorched the grass of the runway. Warsitz reported he could hold the direction of the roll marvelously with the brakes. With a mighty roar the experimental plane with no propeller and no visible engine, unpainted and without markings, lifted off gently. The riggers cheered.

After only one turn around the field Warsitz brought the plane back and landed. He said, "The aircraft is very top-heavy," he said. "Let's make a change." While every pilot prefers to have a top-heavy plane, this aircraft was awkward to handle. The stabilizers were adjusted as Warsitz requested, and he made a few more rounds over the airport at low altitude and landed. The long-awaited day of the first flight had come and this delay had happened. Von Ohain paced back and forth, and Heinkel mopped his brow. All the team was silent.

Again there was the taxi, the high whine, and the roar of takeoff. The onlookers held their breath. At the end of the airfield the plane lifted off. Warsitz went straight off into the distance and the spectators heard nothing more. Five minutes passed, a very long time indeed, before the plane reappeared and circled the airfield.

Warsitz reported: "I kept trying to retract the landing gear, but something wasn't right. The main thing was, she was flying! I chose to fly very slowly, slightly below 370 mph (600 km/h), because I wanted the gas turbine to run at the lowest possible temperature, although it was designed to operate at 373 mph. Herr Schwärzler had told me that on no account was I to go any faster." The aircraft flew no more than 50 miles over the Baltic Sea at altitudes between 2000 and 3000 ft.

Warsitz came closer to the airfield and appeared to line up for landing, but the group on the ground saw him circle again. Von Ohain remembered how he felt: "We said, 'Oh, he feels so good up there that he does not want to come down.' In fact, he could not see the airfield because of the glare of the low early morning sun on the mist, which he later said was terrible. He had to make several circuits before he could safely land. In reality, he was trying to get down through the fog."

The turbine reacted obediently to the throttle lever although, as Warsitz could see from his instruments, a fuel pump had broken. He said, "As I flattened out, I went too far. For fuel reasons, I could not fly another circuit. I decided to side-slip or I would have run into the Warnow River." It was foolhardy side-slipping with a new aircraft, especially very close to the ground—one small mistake, and the wing could touch the ground and the plane would cartwheel and crash.

The aircraft touched down smoothly without bumping. Warsitz taxied to where Heinkel and the other observers were standing. Jubilation broke out. The riggers were waving like madmen. Heinkel and von Ohain were grinning broadly.

In the light of early dawn, Heinkel, with Warsitz beside him, hurried to a telephone and placed a call to Gen. Udet, head of all technical departments of the *Luftwaffe*. Party-loving Udet answered the phone. When they told him excitedly of the success of the jet-powered plane, Udet said, "Everyone is asleep in Berlin," and slammed down the phone.

The riggers lifted Heinkel and Warsitz to their shoulders and all went to the Marienehe dining hall for a champagne breakfast to celebrate. "We had a nice celebration," Von Ohain said. He was toasted and lifted onto the shoulders of his coworkers and shouts of praise went up for the young man. The party was boisterous, described by von Ohain as having "a lot of shouting and jumping up and down." The air was thick with cigar smoke.

The young inventor did not stay at the party. He had his own celebration. After the pictures were taken and the toasts were said, von Ohain took a bottle of champagne from the table and left the room. He went back to the bank of the Warnow River, unfolded a deck chair, and drank the champagne. All that day until after dark that evening he sat alone, looking out over the river.

Warsitz told Heinkel that the flight had been perfect with pleasing characteristics, aside from problems with the undercarriage and the fuel pump. He reported "some instability of the airframe at high speeds, but this caused me less worry than the necessity of flying at low altitude in an unmarked plane over the antiaircraft gun emplacements at the neighboring Arado plant, and over the air base at Warnemunde." Warsitz had reason to be apprehensive because he had been told that the Polish were making raids on Germany at the border, and war was coming.

Von Ohain said of the flight:

> I knew that the flight could at any moment be different. A blade gets loose and BOOM! the whole machine is lost. I was more concerned that Warsitz come back safe and sound. The airplane could later be improved. The pleasing thing was that the pilot felt very good because the aircraft did not have torque as you have with a propeller. We were amused. The pilot had a sense of safety because the gas turbine doesn't shake and doesn't produce a rumble in the cabin that a reciprocal engine produces. There was an absence of vibration and noise like an electric motor, and that psychologically gave the pilot a false sense of security.
>
> We learned some things from that flight. We found a substantiation of the reduction of turbine inlet temperatures at flight, and could therefore be better assured of the calculated flight performance data. We also learned the behavior of the air inlet. We were pleased that our basic considerations were correct. The flight demonstrated not only the feasibility of jet propulsion but also a number of qualitative engine characteristics that had previously been doubted by many opponents who did not believe in the practicability of aircraft gas turbines:
>
> –The combustor had a wide operational range, which made it possible to start the engine easily and to fly at different speeds and altitudes with an accelerating or decelerating engine.
>
> –The combustor could have the above characteristics and yet be made small enough to fit in the engine envelope.
>
> –The flying engine had highly favorable thrust-to-weight ratios.

Von Ohain had created this world-changing piece of technology. He could now say that he had achieved the first step in a new kind of propulsion, flight without the noise and vibration of the reciprocating engine. Flight would now have beauty and elegance.

The world's first flight of a turbojet-propelled aircraft was a success, but there was no announcement of it in the world news, which centered on reports that German troops were massed on the Polish border. Trains and trucks were loaded with supplies. Along the Polish border, civilians were being moved inland. On 28 August 1939 food rationing began for all Germans, beginning with butter, milk, and potatoes. When Heinkel turned on his radio on 1 September 1939, he learned the reason for the Air Ministry's lack of interest in his "little" achievement. Germany had invaded Poland.

Early in the morning of 1 September 1939, five days after the first flight of von Ohain's jet engine plane, all employees of the Heinkel Company assembled in the courtyard of the factory to hear a speech by Hitler broadcast over loudspeakers, announcing the glories of Germany's invasion of Poland. Von Ohain became very quiet when asked about the invasion. He said he would never forget that moment and what went through his head. For him and others in his group it was completely clear that there could be no victory for Germany and that the war would have to be lost in the long run. Germany's aggressions into one country after another would only lead to catastrophe. He recalled, "The loudspeaker continually blared. The contrast was so strange. All our dreams were overshadowed by the war and political happenings."

The entire Air Ministry was asleep to the potential of the dramatic new propulsion system. In the next days after the first flight Heinkel telephoned each day to get Udet or any *Luftwaffe* officials to come to see the jet plane perform. They were evasive. During the months of the Polish campaign, members of the Air Ministry said they had no time and no desire to consider a new form of propulsion. Even without a demonstration they thought the turbojet was an infeasible idea. Heinkel had always had a low opinion of the technical expertise and foresight of Germany's leaders, and their attitude verified it. The war delayed the possibility of rapid development of the new propulsion system.

Two months after the first flight, Heinkel finally convinced Milch, Udet, and Helmut Schelp (Hitler and Göring had other plans) to come to Marienehe. They arrived cold and tight-lipped in attitude, although Heinkel could not discover the reason why.

The demonstration of the He 178 for these dignitaries was nearly a disaster. The aircraft's gasoline pump seized up as the plane was taking off. This account of what happened is from Heinkel's autobiography, *Stormy Life*:

> The jet unit ceased abruptly. Warsitz taxied toward a nearby railroad track, but managed to swerve at the last minute. I sensed a catastrophe that would destroy all my hopes. I jumped into my Mercedes, leaving my visitors behind, and dashed over to Warsitz. By some miracle, the 178 seemed undamaged.
>
> "The plane must fly," I roared. "If those fellows go back to Berlin, everything is finished, absolutely finished."
>
> Warsitz passed his oily hands over his face. "If you take Milch and the others over to the club for a while, we'll put the plane in order."
>
> "Fine," I said. "Now give Milch some harmless explanation for the accident. Think of something. Say whatever occurs to you, but for heaven's sake, think of something."
>
> The faces of the visitors were even icier than before. Milch said, with a hint of a jeer, "Well, what's it all about?" With a straight face, Warsitz said, "A tire blew out."

Heinkel quickly led the dignitaries to his club and provided a lavish champagne breakfast served by pretty girls. He called it his *Verzögerungsfrühstück* (delay breakfast). Two and a half hours later a very calm Heinkel led them out of the club and a perfect takeoff and flight took place. Udet seemed amazed. Heinkel reported, "Warsitz made such a perfect landing that the machine rolled to a stop in front of the group. I felt we had won. My disillusionment came when the visitors, without any further remark, left [the flight line]." The visitors did not seem to see that they had witnessed the beginning of a new era in powered flight.

All of them went to the company clubhouse for a discussion of Heinkel's next project. He had plans ready for a turbojet aircraft based on what they had learned from the He 178. At the meeting, the Air Ministry (RLM) agreed to reimburse Heinkel for all costs of the development work on the jet engine up to that time.

Information about the He 178 was reported in an interview with a prisoner of war taken into custody by the British a year or more after the first flight. He was able to produce three pictures of the He 178 aircraft on the ground: full front, front from above, and back from above. The date of the interview was not recorded. The report was not entered into U.S. government records until 9 April 1946.

Information was, of course, estimated by the observer. It is interesting to compare it with the data from the first flight. The original report was written without breaks. It has been divided into paragraphs for ease of reading.

HEADQUARTERS
FREEMAN FIELD
Air Materiel Command
Seymour, Indiana

FOREIGN EQUIPMENT DESCRIPTIVE BRIEF
Serial No. 46-11
Photographs of Heinkel Fighter

A. Purpose:

1. The purpose of this brief is to publish information and photographs on a German aircraft for further identification. This aircraft is not at Freeman Field.

B. Description:

2. The photographs of a German aircraft were received at Freeman Field. The aircraft is believed to be of Heinkel design. The information contained in this brief was extracted from P/W [prisoner of war] interrogations and is not confirmed as to its accuracy. The Heinkel single seat experimental fighter is referred to as the HET. This aircraft is described as an all-metal mid-wing monoplane of 26.2 ft. span. The length of the fuselage is 19.6 ft. The wing was to have been fitted with a high lift device, possibly a Fowler flap, to increase the wing chord by approximately 2 feet.

The aircraft was intended for high altitude operation and the cabin is pressurized. A turbine of the propulsion unit was stated by a P/W as having been located in the forward part of the fuselage. The compressor is said to be of the exhaust turbine type with a maximum speed of 25,000 to 30,000 rpm. The propulsive "tuyere" is in the tail of the fuselage. The design is generally similar to the Italian jet-powered Caproni-Campini aircraft.

No information has been received as to the designer or manufacturer of the propulsion unit. The power ratings were said to be 110% thrust for 5 minutes and 90% thrust for 30 minutes. Performance figures as stated by P/W seem incredible. The maximum speed was said to be 750 mph at 52,000-53,000 ft. at 95%

> thrust. The fighter was reported to be able to reach 19,500 altitude in 1 1/2 minutes, probably at best climb angle of 30 degrees.
>
> Maneuverability, as might be expected from lack of inherent leverage to the airplane, was said to be unsatisfactory especially in turning or in violent maneuvers, at high speeds.
>
> The aircraft was said to be equipped with under wing rocket fittings for assisted take-off. The airplane was intended to be launched by catapult [...blurred words...] runway 130 ft. long. Landing speed was said to be of reasonable speed, probably about 135 mph.
>
> No indication as to what type of armament this aircraft was to employ has been found. No aircraft answering this plane's description has been located or observed by Allied pilots or by photo reconnaissance. The aircraft was believed to have been used tactically as an interceptor.

Looking back, von Ohain mused, "We came from nothing and flew in only three years." From 1936 to 1939 the turbojet engine was created at the Heinkel Company with the following steps to its first flight:

1) It was presented to Heinkel as a concept on paper in March of 1936.
2) A hydrogen engine demonstrated that it could work.
3) Great difficulties with the combustor were overcome.
4) A special airframe was designed and built.
5) Finally, the successful engine was flown.

Contrast those three years with the time required for a new engine today. At a minimum, 10 years separate a new engine's conception from its manufacture. An operational engine of the year 2000 has been built on "advanced technology" from the late 1980s. The research upon which it is based is much older.

Von Ohain was a physicist. The designers and engineers who solved the engine's problems were not from the world of engine manufacture. Hans philosophized, "Quite frequently, inventions are made by people outside the mainstream altogether. Neither in Britain or Germany did the aero-engine companies think of the jet. They were too busy making piston engines to think about alternatives."

One author has called Hans von Ohain "the man who changed the world." The turbojet plane in its further development has changed our lives immeasurably. Comfortable flights of long distances are attainable in a matter of hours, and fresh food and fruit are available year round. There can be emergency medical transport and disaster aid and advance warning of dangerous storms and military buildups. It has improved the health of the world's economy and brought the people of the world closer together. It provides the safest, fastest, and cheapest transportation for the whole world.

Chapter 9

The next engine and the War

Hans von Ohain and his employer Ernst Heinkel were thwarted from their goal of pursuing the new form of aircraft propulsion by the historic events of World War II.

At the center of each German citizen's world was Adolf Hitler, who projected all-encompassing power. He had a mesmerizing personality and used theatricality to play on the emotions of the German people. Marching bands, swastikas, uniforms, and oaths of allegiance inflamed the enthusiasm of the masses. Large pictures of the *Führer* were displayed everywhere. Hitler used the new media of radio broadcasts and films to keep his image and authority projected into every part of life. The German government's slogans such as *Kraft durch Freude* (Strength through joy) and *Arbeit macht frei* [Work makes (us) free] were inspiring. History records that the slogans became negative when the weak were persecuted and the *arbeit* (work) was in a concentration camp. Hitler's officers took a "flag oath" pledging to defend Germany at all costs, words that sustained their patriotism even in later years when they doubted their leader.

Hitler's National Socialist Party had risen to power early in the 1930s when von Ohain was a student at Göttingen University. The Party's strength lay in its philosophy as an alternative to the much-feared communism. Hitler increased employment, a very welcome change after Germany's period of deep economic depression in the late 1920s, by sponsoring many public works. He directed large government expenditures to rearm the country. People had enthusiasm for the good that Hitler and the Party appeared to have done for the country. There was renewed pride in the "Fatherland" that had disappeared with the defeat in World War I. When Hitler annexed the Rhineland-Saar and Sudetenland, it seemed the inalienable right of Germany. With the takeover of Poland in 1939, propaganda made that action appear to be almost right, but the wisest men began to question Hitler's ambitions.

Hans von Ohain had been affected very little by World War I during his childhood because his family was staying with his grandparents in the quiet city of Magdeburg, while his father was away from home serving in the military. As he became a teenager, he could see the effects of the great economic difficulties caused by reparations following the war, and he began to see changes caused by the turmoil of Germany's political factions as he completed his *gymnasium* education. Von Ohain was studying at Göttingen during the early years of the National Socialist Party takeover and kept himself isolated from political interference. With the beginning of Germany's aggressions against neighboring countries, von Ohain was grateful that he was a part of Heinkel's organization, which kept him very busy and isolated from many events in the local community.

Heinkel's next turbojet aircraft project, the He 280, began even before the first flight of the He 178. Data from the He 178 airframe and He S 3B engine provided the basis.

The design of the airframe and engine required long review and planning sessions involving Hans von Ohain, Heinkel's new technical director, Robert Lusser (Lusser had previously been at Messerschmitt), and Heinkel's project office. The He 180 airframe, prototype of the He 280, had been fabricated by Heinkel designers by June 1939. The He 280 airframe was intended to be a production aircraft. The earlier configuration with a single fuselage-mounted turbojet was abandoned in favor of a twin-engine design for additional thrust, with engines externally mounted under the wings. The external engine installation required the smaller engine diameter of an axial engine. It was designed to use the axial-flow He S 30 engine, the Wagner engine that Müller's team had brought with them from Junkers Airframe.

In November 1939, on the day of the demonstration of the He 178 for Udet and others of the Air Ministry, Heinkel showed the plans for his next jet aircraft. Heinkel proposed not only a new aircraft but also the dramatic innovation of a new way to produce aircraft. He proposed engine-airframe integration at his own factory, pairing a Heinkel airframe with a Heinkel engine. Helmut Schelp was in favor of this plan, but Mauch continued to want engines produced only by engine manufacturers. In November 1939, Heinkel received a contract to construct the He 280 aircraft. At the same time, the Air Ministry awarded a contract to Messerschmitt to build the Me 262 airframe.

In 1940, Heinkel learned that a factory in Stuttgart that made small engines, *Hirth Motoren* (Hirth Aircraft Engine Corporation), was for sale because of the death of its owner, Hellmuth Hirth. Heinkel saw an opportunity to expand his business into a facility with capabilities already in place to build engines. He wrote to Udet at the Air Ministry on 11 June 1940 with persuasive arguments for intercession on his behalf. He cited the value of getting jet engines into production while avoiding a large capital investment considering the war effort, the suitability of the Hirth company, and a sentimental argument that he could continue the life's work of one of his closest friends, Hirth. The ambitious Heinkel had four projects contemplated for the facility, not only turbojet engines.

> Acquisitions General
> General of the Air Force Udet
> Leipzigerstrasse 5-7
> Berlin W.8
> Subject: Wish to Purchase Hirth Motors Ltd.
>
> Dear General:
>
> It is well known that the development of jet engines has proceeded separately from airframe manufacture. This historical development has fully proven itself in practice because the influence of the airframe shows up more strongly for the high speeds associated with jet engines than for normal engines. The technical status of our jet engine development has advanced so far today that we have to start thinking about how to produce them. They have shown so much potential for applications.
>
> Because we are conducting our development along several paths, four different types at the moment, we also need an expansion of our fabrication facilities. This expansion would be possible by building a new factory, but this would require a large capital investment which should be avoided at the present moment. We believe that merging with an existing engine manufacturer is preferable. Our deliberations have led us to the view that Hirth Motors Ltd would be especially suited for this purpose.
>
> The Hirth firm is already working in areas which are completely related to jet engines, turbochargers! The manufacturing facilities of the Hirth Co. are tailored to air-cooled engines of small displacement. This fact permits a favorable connection with our "ducted propeller" engine, currently in development, for which an air-cooled, two-cycle engine is planned. We would like to note also that development on this project has reached

> the stage that the engine is also applicable for normal "propeller propulsion."
>
> As we have heard, several firms are interested in acquiring Hirth Motors. We believe that some of these already have such a large manufacturing capacity that their expansion is not so urgent as it is for us. Others have not done anything in this special technical area.
>
> Finally, I would like to remark that I, as a Wurttemberger and one of the oldest and closest friends of [the late] Hellmuth Hirth would be especially pleased to take over and continue his life's work. I believe that my personal interest and the proven organization of my firm will relieve the emptiness that probably burdens Hirth today. I also have good connections with the state government of Wurttemberg, so I believe they would be pleased if I took over the company.
>
> Heil Hitler
> Your very devoted
> /s/ Ernst Heinkel

Udet read and remembered. Udet was not an advocate for this new turbojet engine, but he stood by Heinkel when there was constant pressure from Mauch to turn Heinkel's engine projects over to some engine manufacturer. Udet gave Heinkel a verbal promise: If flight testing for the He 280 began no later than April 1941, Udet himself would make it possible for Heinkel to buy *Hirth Motoren*. There were only 16 months between the contract date and the target date of completion, April 1941, for both development and ground testing of the engine.

Hitler's war was proceeding easily. Troops marched into Norway and the Low Countries. The occupation of France followed, considered by the Germans to be a successful coup against an old adversary. Little need for long-range planning seemed to exist. Udet's office at the Air Ministry issued the edict banning all aircraft design that would be completed more than six months in the future. The April date proposed to Heinkel exceeded that deadline by some weeks. Udet waved away Heinkel's concern and repeated the April challenge for continuing development of the He 280.

An axial engine was considered to be the best power source for Heinkel's new jet aircraft, but it was a more difficult design challenge than the centrifugal type engine. The Heinkel Company had investigated axial engines earlier. In 1938, on the recommendation of von Ohain, Heinkel had placed Walter Encke of the AVA in Göttingen under contract as a consultant. Encke had presented papers on single stage axial compressor studies. He advised Heinkel that the construction of a multistage axial compressor would offer high total pressure and high efficiency and would be possible but speculative with the technology available at that time. Shortly afterward, Encke was hired by Junkers and assisted with the design of the eight-stage compressor of the Jumo -004 engine. Instead of exploiting the simpler centrifugal type engine, as in the British and U.S. turbojet programs, the Germans plunged headlong into the more advanced technical requirements of the axial flow engine. The advantage they held in the early successes with centrifugal engines was lost. At the Air Ministry, Hans Antz and Schelp drew up specifications for an axial turbojet, and von Ohain began a design that was laid aside due to the urgency of his other work.

Heinkel owned the von Ohain He S 3B centrifugal engine. Because he wanted to be the jet propulsion industry leader, he had hired Max Adolf Müller and his team when they left Junkers in order to acquire the He S 30 Wagner-designed axial engine. It was, in concept, the most advanced axial engine design of its time—compact and lightweight, but difficult to build. At the time his team was hired, Müller had given the impression that the engine program was far ahead in development. It quickly became apparent to von Ohain and Heinkel that this program was not at all at the development stage first described, and the He S 30 engine would not be ready to meet Udet's challenge. The engine had never run on its own power, only when it was hooked up to an electric motor.

Müller was in charge of the He S 30 engine program. He was a very talented engineer but did not work well with other people. Müller demanded preferential treatment from Heinkel for what he considered to be his outstanding technical contributions. He made a number of frustrating empty promises. He did not like to work under pressure and was unable to make sufficient progress with the engine. Heinkel told him, "Of all my 50,000 employees, you are the most trouble!" Müller was allowed to resign his employment. Heinkel and his technical director, von Ohain, turned the He S 30 engine over to Rudolf Friedrich.

Meanwhile, the German euphoria for Hitler's successes vanished with the invasion of the Soviet Union. Every German student had learned about Napoleon's calamitous Russian campaign. Assurances from the government that the Russian campaign would be over in six weeks were met with doubt, especially among the military. The German generals appeared to lack the courage of their own convictions to stand up to Hitler. Reports of death and suffering on the Russian front came back to a shocked German populace. The young boys who were soldiers at the front lost any ambitions to be heroes. They laid aside any idea that they were fighting for Hitler and instead fought because it was their duty for the Fatherland. Von Ohain was upset to hear that his brother, Wolf, was inducted into the army immediately after completion of his *gymnasium* education.

At home, German people had sufficient food, although somewhat lacking in variety, because the government had a rationing plan and an efficient distribution system. Most of the people who owned cars contended with the surrender of personal transportation because car tires and batteries had to be turned in to the government in 1940. They had to rely on Germany's vast network of public transportation. Von Ohain, as a valued government scientist, was able to keep his car for transportation to work.

The problem of developing an engine for the He 280 aircraft was now delegated to von Ohain. His gamble was to create an engine that could meet the deadline for the first flight by April 1941. There was no time for mistakes. He decided his best option would be to stay close to the successful He S 3B engine design from the He 178 flight. He would introduce only those changes that were absolutely necessary for significantly reducing the engine diameter because the diameter of the He S 3B engine would be far too large for the airframe of the He 280. The front combustor of the He S 3B engine was abandoned and the combustion chamber was put between the compressor exit stator and the turbine inlet stator. These design changes resulted in an engine configuration designated by Heinkel as the He S 8A.

Modifications and improvements of the He 280 airframe were made while the engine was being developed. The He 280 had an exceptionally clean fuselage with an overall length of 34 ft, 1 in. Stressed skin metal construction was employed for the airframe, with an oval-section monocoque fuselage and twin endplate vertical tail surfaces. The elliptical wings spanned 39 ft, 5 in. Unquestionably, it was the most advanced warplane of the time. The airframe had revolutionary design features, a tricycle undercarriage, and a compressed air ejection seat. The tricycle undercarriage permitted the aircraft to taxi in horizontal position, avoiding the problem with jet blast that arose later with the prototype Me 262, which retained the conventional tail-wheel undercarriage. It was the first aircraft to be fitted with an ejection seat that was used in later tests, giving the plane the distinction of being the first to use such a device. Provision was made in the original design for the eventual introduction of cabin pressurization.

Two of von Ohain's He S 8A centrifugal-type turbojet engines were slung under the wings of the aircraft, their thrust lines passing beneath the raised twin fin-and-rudder tail assembly. Each engine was rated at 1300 lbs (5.9 kN) of static thrust. The radial inflow turbine was a concept representing a technology leap at the time, but its drawback was the requirement for nickel-iron alloys for the high temperatures, and nickel was in short supply. It was expected that with further development the He S 8A turbojet would eventually offer 1580 lbs (7 kN) of static thrust and a maximum speed of 560 mph (900 km/h). The Heinkel team predicted with reasonable confidence that the engine would have a lifetime of up to 10 hours. Ground testing of the He S 8A began early in 1941, but the engine did not produce the required thrust or sufficient reliability. Von Ohain's group was plagued for some time with turbine blade fatigue failures near the blade roots at the turbine exit section. As the April 1941 deadline grew closer, they spent long hours adjusting the engine for the airframe. They were uncertain when the test flight might take place.

The world's first jet fighter and the first twin-engine jet-powered aircraft, the He 280 with He S 8 engines, made its flight on 30 March 1941, two days before the April deadline. It gave Heinkel and von Ohain another historic first. In the *Royal Air Force Flying Review* of November 1958, Fritz Schaefer, the plane's test pilot, wrote:

> During the drive to the airport, Dr. Heinkel had advised me not to fly the prototype at this stage, and to limit my initial test to high-speed taxiing and, possibly, a short "hop" on the runway. But as he continually praised the new fighter and went to considerable pains to remind me that Erich Warsitz had test-flown the very primitive He 178 research aircraft nearly two years before, I could see that his admonition not to take the He 280 V 2 into the air was not entirely serious.
>
> The runway at Rostock-Marienehe was little more than a thousand yards in length, extending in an east-west direction from the Warnow estuary to the rampart of the Rostock-Warnemunde railway, and with a fast taxi and a short "hop" I might well overrun the strict limitations of the runway! As if he had read my thoughts, Heinkel turned to me as I left the car and said: "Do as you like!"

> The first flight was obviously to be a matter of some delicacy. The He S 8 turbojets had never been tested in the air and had, in fact, only just completed one hour of bench testing. The cowlings had been left off as it had been discovered that quantities of gasoline tended to collect in the bottoms of these while the turbojets were running, presenting a very serious fire hazard.
>
> The He 280 V 2 was not entirely new to me, of course, as I had flown the machine as a glider on several occasions in order to gather data on the general characteristics of the aircraft in the air. For these tests dummy engine nacelles had been fitted and the prototype had been towed to altitude by a Heinkel He 111 bomber.
>
> I hoisted myself into the cockpit; switched on the fuel. The turbojets came to life with their characteristic high-pitched whine. A hasty cockpit check, brakes off, and I was rolling rapidly down the runway.
>
> My main thought was to conserve the mere one hundred gallons of fuel in my tanks—just enough for a single flight around the airfield. The greatest difficulty was the fact that the primitive turbojets had no automatic regulating devices, the throttle cock being the only means of controlling the thrust of the power plants. I had to watch the engine rev. counters extremely closely as it was dangerous to exceed 13,000 rpm even by two or three hundred revs, and below this figure the turbojets did not develop sufficient thrust to enable me to take off.
>
> The rev. counters rapidly indicated 13,000 rpm, although by the time I had reached the runway's halfway mark they had dropped to 12,500 rpm. They quickly picked up, however, and I was airborne—but without much runway to spare. I climbed to 900 feet and throttled back to 12,000 rpm beginning my circuit. There was no time to raise the undercarriage because the red warning lights were blinking, indicating I was already flying on reserve fuel. I turned over the Warnow [River], made a fairly fast approach, but landed comfortably on the runway with 40 degrees of flap.
>
> The world's first jet fighter had flown successfully!

Although the flight lasted only three minutes, they had achieved the contractual date for flight. Heinkel believed that the potential of the He 280 had been proven in the demonstration.

Six days after its first flight, the jet-powered He 280 was demonstrated for Gen. Udet and others of the Air Ministry. They were impressed by the performance, but they again questioned the need for "such unorthodox warplanes." A mock combat between the He 280 and a contemporary Focke Wulf (Fw 190A) propeller fighter was arranged for them to see. The He 280 displayed clear superiority. As a result the Air Ministry sanctioned preparations for the manufacture of 13 He 280A-O preproduction examples of the Heinkel fighter. The test flight had been an outstanding display, but there were continued shortcomings with both the airframe and the engine. The He S 8B had problems even with a new compressor and modified turbine. The accessory unit needed a better oil pump. Fuel nozzles persisted in becoming dirty and sooty. The aft bearing tended to overheat. The team worked on modifications, but no significant increase in thrust was achieved.

British Intelligence was alive in Germany. The following report, here excerpted, was not the only intelligence report made to Allied intelligence offices about the He 280, but it shows that the German Air Ministry and Heinkel were doing an outstanding job of keeping jet development and the He 280 well hidden.

In September 1942 the report, then marked "SECRET," was given to the Air Ministry, Whitehall, London. The report is interesting to compare with what is known of Heinkel's schedule, the work of von Ohain and his team, and the capabilities of the aircraft and engine already demonstrated.

Foreign Aircraft Bulletin No. 6, A German Jet Propelled Aircraft
[Note: Italics are that of the author, for emphasis.]

> Reports have been received from time to time of the construction by Heinkel of aircraft powered by jet propulsion units. The people engaged in this development *are all members of the Party* and the details of the designs and the stages of their development are considered to be closely guarded secrets.
>
> The enthusiasts who were engaged in the work about two years ago had hopes of producing operational aircraft by 1943 although the more disinterested *members of Heinkel's design staff seemed to consider that no useful results could be shown before 1945.*
>
> The design on which the main hopes were founded was given the designation He 280, but there is, however, as yet, no

report from observers who might have seen it flying and no useful indication as to the present stage of development.

The aircraft, as designed, was to be powered with two thermodynamic turbine units, said *to be based on the researches of the Italian, Campini.* Air from the atmosphere is compressed by a turbo-driven compressor and liquid fuel is burnt in conjunction with it. The products of combustion are exhausted into the atmosphere through a jet to provide the propulsive thrust. Much of Campini's research work has been published and is available in translation. A patent was published by Heinkel (Serial No. 276-572) and described in "Flight" for Feb.19/1942, "Aircraft Engineering" for February 1942, and other publications, see Appendix "A." Apart from this pointer, the details of the power unit are still a German secret.

Following this passage is a description with dimensions as estimated by the reporter. The retractable undercarriage is noted with the comment "possibly tricycle." It continues:

The power units are on the wings, about 1$^{1/2}$ to 2 metres (4$^{1/2}$ to 6$^{1/2}$ feet) to the right and left of the fuselage; they are said to be "just like nacelles but with a long tail."

The performance was estimated by the designers to be:

Maximum speed	800 to 850 km/hr. (495 to 530 mph)
Maximum ceiling	about 15,000 metres (49,300 ft)
Rate of climb with full load	30 metres/sec. (5,900 ft/m)

It was further claimed that the aircraft could climb almost vertically at a speed of 650 km/h (404 mph = 35,500 ft/min).

Range and endurance: At a height of 6,000 metres to 10,000 metres (20,000 to 30,000 ft), a range of 1500 km; 800 km/h (930 miles at 495 mph) at full load was claimed, and an endurance, "at a high speed," of two hours.

Without recopying the entire document these are other comments in the report:

Cockpit: To enable the pilot to make a parachute jump if necessary, a special system is said to be arranged, i.e., sprung or compressed air cushion which would eject the pilot complete

> with parachute and seated on the seat. This innovation is said to be necessary in view of the high speed of the aircraft.
>
> No pressure cabin was contemplated at first in view of the short range and endurance but the question of increasing the range has been studied and it is understood that later developments will probably involve a pressure cabin.
>
> General: Flight at reduced speed is *said to be impossible because the power output cannot be adequately regulated* and the airbrakes are not very efficient.
>
> Developments: There has been some discussion in German aeronautical circles as to how this aircraft is to be used for war purposes. *Owing to its lack of manoeuvrability it would be a poor fighter* and its range and carrying capacity do not fit it for bombing. It is considered to be too expensive to be utilized to ram a bomber. It has been suggested that its main function at first would be a "demoralizing" aircraft, for tactical and scare purposes, similar to the use of Big Bertha in the last war.

The report ends with the following statement:

> To enable flights at lower cruising speed to be carried out, *it has seriously been proposed to install an auxiliary engine and airscrew of orthodox type.*

The British Air Ministry's reaction to the description of this aircraft would be interesting to know. The possibilities of this German fighter had to be impressive although the threat did not seem to be imminent. There is no mention by Sir Frank Whittle in his autobiography that he was asked to verify that such an aircraft with the described propulsion system might be possible.

The timely flight of the He 280 aircraft met Udet's challenge and laid the groundwork for Heinkel to buy *Hirth Motoren* in Stuttgart in the suburb of Zuffenhausen. The Hirth company built small aircraft engines for general aviation and turbo superchargers for large engines. With the Heinkel-Hirth Company, Ernst Heinkel finally achieved the status of owning a fully recognized company for aircraft gas turbine development and manufacturing. Von Ohain at last would have a facility equipped to build engines. The sale was made over protests from other manufacturers, especially Messerschmitt, who delayed Heinkel's acquisition by several months. Udet had been persuasive in the Air Ministry decision to allow Heinkel to purchase the factory. Udet believed in Heinkel's managerial ability and outstanding leadership, and in the technical competence of the 31-year-old Hans von Ohain.

Heinkel moved his engine work, first the He S 30 and later the He S 8 and other prototype engines, from the Rostock area to the new location. Von Ohain continued to be head of all jet engine technical development. The integration of the former Heinkel team with leading engineers of the former Hirth Corporation proceeded very smoothly under plant manager Curt Schif. Von Ohain found an almost unbelievable change from the mode of operation of previous years. The new team had the strength of a sufficient number of employees and top engineering talents, with highly experienced and dependable engineers, technicians, and support personnel, and excellent facilities. For example, Max Bentele, a recognized expert in blade vibrations, was chief engineer. Bentele was placed in charge of component development and testing, including a 6000-horsepower compressor research facility.

Bentele once gave this "roast" of his association with von Ohain:

> There was little we knew about the outfit [the group from Heinkel at Rostock]. One of his aeroengine groups had made the world's first jet-propelled flight. Its leader had the awe-inspiring name of Herr Dr. Phil. Hans Joachim Pabst von Ohain, and this guy was desperately trying to develop the most powerful and advanced jet engine. Quite an ambition, because he was not an engineer but a mere physicist and as such of a lower calibre than his landsmen: The quantum Max Planck or the atom-split Otto Hahn or the relativity Albert Einstein. That was all our "G2" [spy] could find out.
>
> When we learned, first from the grapevine and then through official channels, that von Ohain and his group would join us in Zuffenhausen permanently, we square Swabians felt that cooperation with this Prussian *Junker* will be bloody murder, and it almost was. Meeting him in person for the first time we were somewhat relieved. His boyish looks and behavior gave us the confidence we could handle him all right and mold him to our liking.
>
> The North (Heinkel) and South (Hirth) team were introduced to each other. As in the Serbian saying, "Every gypsy praises his horse," we felt proud of our tradition, or expertise in turbochargers, blade vibrations and mechanical design, and they,

of course, knew everything about jets. It was decided to keep the two groups segregated but equal and von Ohain and his group would ask us for help, if and when they felt they could benefit from it. In view of their self-acclaimed superiority and generally recognized arrogance that call for help took a long, long time to come—about a week. Then the two groups were integrated, they doing the theoretical work and we the practical one.

Our suspicion that all people north of the Limes [a Roman wall built across Germany to keep the unruly barbarians of the north from invading] still lived in a primitive world was soon confirmed. A typical example was their approach to the aerothermodynamic design of a compressor. They called it a "blower," a term traditionally not used for the heart of a jet engine. They first drew simple enthalpy/entropy diagrams, geometric flow areas with streamlines, velocity triangles, and blade profiles. Finally, they had the overly optimistic idea that the air would follow their designs and predictions, and thus testing such a contrivance would be a mere formality.

One morning after my test crew had put an experimental compressor through the paces the night before, Dr. von Ohain and his aerodynamic cohort came to my office to hear the good news first-hand. With my typical Swabian frankness I said, "Sure, we tested the beast thoroughly by running and measuring the full ranges of speed, airflow, and pressure ratio. It's still in one piece because we had done the mechanical design, but if you think its performance was anywhere near your predictions and expectations you must be utterly naïve or have rocks in your head."

This well-meant remark stunned both. They got red faces, mumbled something about lack of cooperation, antipathy against intellectuals in general and Prussians in particular, hurriedly left the room and slammed the door in my face. I don't know who of the two had actually banged the door. I feel Hans's nobility would subconsciously prevent such a blatant breach of etiquette, but I am not sure.

We got even with the full jet engine. She hit the thermal barrier 350 kilograms short of her 1,300 kilogram design thrust. As long as they debated the effects of the specific speed of the compressor and turbine, forced vs free-vortex design of the turbine blades, etc., we intuitively felt the turbine needed more breathing space. Clandestinely we took the turbine wheel off

> the engine, de-bladed it, clamped the blade roots on a vise, and twisted with a crowbar the blade tips 12 degrees open. That did the trick! Engine thrust came up to par. Theoreticians and practitioners were on equal footing and we became real friends – almost.
>
> We were still *verboten* to have lunch with von Ohain in the Executive Dining Room, but we could sit with him on the same bench in the air raid shelter. There we had plenty of time to debate our fate, namely, to be overrun by the armies of the French, British, Americans, or, God forbid, the Russians.

Bentele says of his association with von Ohain, "I did not know how it would be to work with this boy, but within a short time we established a close relationship as well as a frank rapport. Technically he became my immediate superior though I never had the feeling that he acted as my boss. He favorably differed from the Prussian image so much disdained by Swabians." He praised von Ohain: "I recognized his exemplary character, his absolute honesty, his extreme fairness in judging his coworkers and others, and his love for associates who pulled their weight. His charm, modesty, and humility were engaging. In short, we were on the same wavelength; we 'clicked.'"

Von Ohain was very much impressed with Bentele's reputation and work and commented, with a smile, on his very strong personality. "One lets a person like this take charge," he said.

Von Ohain's particular strength of leadership was to let a person express his ideas and then support him to let the ideas come into use. Von Ohain worked with each person to help him reach his goals, interjecting his own ideas where assistance was needed. This management approach was a surprise to many who had never had such an opportunity before. Von Ohain not only earned the undying respect of those he worked with, but he also gained the benefits of many good ideas that may not have been offered in other circumstances.

Before 7 December 1941, the United States had been observing Germany's aggressive moves across Europe, particularly the bombings in Britain, and felt sympathy for the people who suffered through the attacks. Then the United States was attacked by Japan at Pearl Harbor. A declaration of war toward both Germany and Japan followed, joining the United States with Britain, Free France, Russia, and other allies.

Allied bombing now brought destruction to the German homeland, with long years of difficulties for the people. Life through the war years is seldom discussed by older Germans today. The youth of Germany call it "the old people's war." Survival of World War II is a silent misery shared only with others who were there.

Long queues for bread and other basic necessities took the housewives' time. Small gardens were planted where there was space in the cities, and city residents went to the country to barter with farmers for food. Instead of the natural abundance of Germany, food was replaced by the ersatz: sawdust-laced sausages and bread, imitation sugar, coloring added to canned vegetables to make them look palatable. Health began to be affected by prolonged hunger and problems from food additives or inadequate diet. Slow starvation showed in sallow faces and red-rimmed eyes. Cancer and other systemic illnesses were prevalent. Coal for home heating and fuel for transportation became more and more scarce. The deep snows of winter added to the suffering. Those who had to work rose out of bed into bitterly cold rooms and walked long miles to get to a factory. Walking to work through the cold and snow grew worse with shoes that were wearing out, with replacements not available.

The government instituted clothing collections and citizens were considered unpatriotic if they did not relinquish a second coat or pair of shoes to share with others. Factories began extended working hours and were staffed by the old and very old, workers too old to be eligible for uniformed service. Maintenance of public transportation remained a priority, and after bombings the tracks and streets were repaired remarkably fast. Repeated bombings by the Allies meant loss of shelter and perhaps loss of everything. People in bombed-out cities lived in basements in terrible conditions. Every night sleep was interrupted, and citizens went to air raid shelters. The fear of betrayals to authorities was constant, and in some sad cases even family members reported each other to authorities because of a comment against the government. As a whole, the civilians tried to care for each other and shared. Many opened their homes to others who had lost their homes in bombings or had come from other areas where fighting was intense.

The German population's knowledge of concentration camps is always questioned. The policy of harassment of Jews started in the early 1930s. It was not until February 1942 that the government began the first actual deportation of Jews to concentration camps, and Germans believed them to be work camps. Rumors of other activity at these camps seemed incomprehensible and were disallowed. As one said, "You didn't believe it, you didn't want to believe it, you just didn't think it was possible!" If those living near the work camps had doubts about what was happening, they were afraid to speak out. The German people worried greatly about their sons and husbands at the front and grieved for those who had been lost in conflict. They lived daily with food shortages and cold and air raids. There was little time spent on worry about others outside of their own families.

The brilliant scientist Theodor von Kármán made an angry statement about concentration camps: "That an intelligent man did not know what was going on in his own country was beyond my belief. Some people found it convenient and comfortable not to listen. I don't think there were any serious efforts made to find out the truth."

Leaders of the Nazi Party lived very differently from the rest of the German people. They luxuriated in houses relinquished by Jews, with servants selected from slave labor from Poland and Czechoslovakia. They had enough to eat, fine wines from France, and oil paintings from "liberated" museums.

As the war progressed and retributions began by Russia, Germans feared the Russians more than any action by the Allies from the West. As the German army fell back from positions in Russia and Poland and Czechoslovakia, many civilians abandoned their homes and farms and fled westward ahead of the invading Russian army. Those who stayed lost their fathers and sons to camps in Siberia. Women who remained tell of living through events more terrible than death.

Von Ohain told little about living through the war. Characteristically he was a person who did not look back, but to the future. When pressed for a comment he said, "I was busy. I was working." Von Ohain was a peace-loving man. He believed that the war was lost by Germany from the very beginning. Von Ohain had the comfort of working for Ernst Heinkel, who counted on the work of his staff and did his best to care for them. Nazi Party membership was not mandatory unless one had an official position, and Heinkel protected his workers, including von Ohain, from needing to join the Party.

On one trip to Berlin, von Ohain was riding a streetcar when a "nice-looking, very friendly man" got on. Sewn on his coat was the big yellow Star of David. Von Ohain thought, "Oh, that is a Jew." His immediate reaction was, "I felt so unbelievably ashamed of being a German." He closed his eyes and remembered, "I still see it. I never forgot it, never, never! That was almost the worst thing I ever experienced, and it was in Berlin near my home."

Von Ohain was asked if he had worked with any Jews at the Heinkel Company. He said, "I don't believe there was one."

Von Ohain had to endure the interruptions of work at the Heinkel factory. Hitler continued his nationalistic harangues and made long speeches broadcast into all the factories in Germany. All work had to stop, and workers were expected to listen attentively to what was said. Von Ohain recalled that he was warned by his supervisor that his facial expression during the speeches was not appropriate. "You look like you are eating sandpaper. This could get you into trouble with party members who are watching everyone." He learned to have a blank expression and to make no comment that could be interpreted against him.

Von Ohain did not tell his family and friends in the United States about the hunger and cold he suffered during the war. He mentioned that after the factory at Stuttgart was bombed he had to sleep in a barn on the outskirts of the town.

Many individuals did keep their integrity in spite of Hitler. Von Ohain was one who found it possible to behave decently toward his fellow humans.

Von Ohain was faced with the increasing complexity of engines, particularly problems with combustion. All materials for building engines were strictly allocated and became scarcer as the war continued. Metal parts were reworked and recast from scrap. The He S 30 axial engine continued in development and made its first run in April 1942. Von Ohain, in his customary mode of giving recognition to a team for good work, was quick to praise it. He described the engine as outstanding: "The continuous thrust was about 1,650 lb. This engine had the best ratio of thrust to weight in comparison to all other contemporary engines. This was the result of its advanced 50% reaction axial flow compressor designed by Dr. Friedrich." The von Ohain He S 8 engines were rebuilt in many iterations, but deficiencies continued well into 1942, with long days and long weeks of small triumphs and many frustrating delays.

Chapter 10

War planes

The German Air Ministry wanted to place a production contract for a jet aircraft. The two options were the Heinkel He 280 with the von Ohain He S 8A engine and the Messerschmitt Me 262 with a Jumo 004 engine. The Messerschmitt Me 262 was chosen. It became the legendary aircraft of World War II, the first operational jet plane with the first mass-produced turbojet engine in the world.

It was called the *Schwalbe* (Swallow) for the fighter version and *Sturmvogel* (Stormbird) for its bomber version. Woldemar Voigt, the airframe designer, met the challenge of matching a high-speed airframe with the novel characteristics of the turbojet engine. It was a very advanced aerodynamic design with twin jets, swept wings, and heavy armament, and a top speed of 540 mph. The manufacturing cost of the Jumo 004 jet engine was about one-fifth of that of a propeller-piston engine of equivalent power output; it required only 700 man-hours to build, in contrast to the 3000–5000 hours needed to build a conventional piston engine. It used axial flow turbomachinery and straight throughflow combustors, and incorporated a variable thrust nozzle. Even afterburning was demonstrated but was never put into production. Later, when nickel was in short supply, the engine used air-cooled hollow turbine blades made of sheet metal.

The Me 262's first flight had been made on 25 March 1942 with BMW P3302 engines. Just after takeoff both engines failed, but the aircraft was saved by the front-mounted piston engine that had been added because of the uncertainty of the new type of engine. The next flight was successful on 18 July 1942, more than a year after the first flight of Heinkel's He 280, with two Jumo 004A pure turbojet engines designed by Anselm Franz and his team at Junkers. The Air Ministry ordered production of the Me 262, including 80 Jumo 004 engines. The first production units of the Me 262 built under the government contract flew in the spring of 1944, achieving 624 mph.

A flight of the Me 262 was attempted with two of Hans von Ohain's He S 8A engines, but the engines did not provide enough thrust.

The Air Ministry's decision for the Me 262 was a blow to the jet program at Heinkel-Hirth. Von Ohain had high praise for the Messerschmitt plane: "The reputation of Junkers as a very capable and reliable engine manufacturer was an important factor in the final selection." He continued, "I don't believe it was a loss for Heinkel. Large scale manufacturing of turbojet engines in his factories would not have been possible at that time." Von Ohain looked back on the rejection of his He S 8 series of engines: "If only I had made the engine slightly larger in diameter, it would have been very successful."

The Heinkel-Hirth organization was a mixed success. Heinkel and his company had an outstanding reputation in Germany with a series of "firsts": 1) He 176, the first rocket plane; 2) He 178, the first jet-powered aircraft; and 3) He 280, the first jet fighter

However, in spite of all his money, Ernst Heinkel was not equipped to perform engine production as were the long-time engine companies. He did not have manufacturing equipment, test stands for compressors and turbines, development and design engineers, or quality control specialists. Heinkel's staff at Heinkel-Hirth, about 50 designers and engineers, were the best in Germany for research, but Heinkel's engineers lacked the experience for practical development. Besides lacking production ability, Heinkel failed to have a unified management effort to bring a single project to production status from the many projects he had in development at the same time. It was said that a visitor to the plant would have been surprised that there was a war taking place in Germany because most of the work at Heinkel's facility was pure research. An atmosphere of great creativity prevailed. As leader of the jet propulsion division, von Ohain had every advantage to be able to conceive advanced engines.

Fortunately for von Ohain, Heinkel's inability to go into production kept the reputation of von Ohain spotlessly clean of any involvement with the use of slave labor, contrasting with Wernher von Braun, who received bad publicity when he was recruited to the United States for rocket research.

By the fall of 1942, the extreme shortage of critical materials in war-ravaged Germany forced a decision by the Air Ministry. The number of engines in development was limited. The Air Ministry considered that the Jumo 004 and BMW 003 engines were simpler and easier to produce than either of Heinkel's principal engines. The

He S 8 and He S 30 were cancelled. On 27 March 1943, Heinkel was instructed by the Air Ministry to abandon all further development of his He 280 aircraft as a fighter, continuing with only a few prototypes for testing purposes. The airframe was tested as a glider toward the end of the war when fuel became scarce. Germany frequently used gliders to transport troops as an alternative to deploying paratroopers. Perhaps Heinkel's grating personality caused some personal prejudice against his company by individuals in the Air Ministry. He had lost the support of his friend, Gen. Udet, who had committed suicide in late 1941. Just when Heinkel had reached his goal to have his own aircraft engine company, he lost the authority to build both turbojet engines and jet aircraft.

If the Air Ministry had not demanded that only engine manufacturers produce turbine engines, creative and innovative work might have resulted from new sources. If the Air Ministry had chosen the He 280 for a contract rather than the Me 262, which was a year behind the –280, Heinkel and von Ohain might have overcome its problems and had a jet aircraft ready for production earlier than the –262. If Heinkel, the experienced aircraft builder, had received the Air Ministry contract for the He 280, could he have arrived at production capability for engines? If jet aircraft had been in production in Germany earlier, would the course of the war changed?

Allied air forces had their first indication that the Germans had some new propulsion system early in the war. Puzzling reconnaissance photographs showed that some German runways had scorch marks along the full length. Frank Whittle recalled being shown photographs and guessing that they were made by turbojets.

The Germans' turbojet-powered planes were a frightening weapon to the Allies. The speed exceeded anything they could imagine. Out of nowhere a turbojet plane could appear unexpectedly, attack, and go out of sight before the Allies' slower planes could react. A U.S. Army Air Corps cameraman made an impressive film of such a sighting that is played frequently at gatherings of air enthusiasts. In the film, the jet plane flashes past the prop plane and you hear the shout from the cockpit, "What the hell was that?!" (It is remarkable if a cameraman with sound recording capability was sitting in the cockpit at that exact moment. Sound may have been added later.)

Because the jets flew relatively slowly during takeoffs and landings, they were particularly vulnerable to Allied fighter attacks near their airfields. Attacking the German jets when they were low-and-slow became a successful defensive strategy for Allied pilots.

The Me 262 jet fighter, along with the V-2 rocket, is remembered as one of the Third Reich's "wonder weapons" that presaged military weapons of the future. Squire L. Brown, formerly chief of the Technology Assessment Division at Wright Laboratory, made these comments:

> The Me 262's speed could evade the Allied fighter escorts, but made attacking a bomber difficult. It had only a few seconds of opportunity to fire its rounds, and the firing rates of its armaments were too slow to get off the many rounds needed to do lethal damage to a bomber. A B-17 was slow, seldom exceeding 200 mph, and was a tough target, since only a hit in the cockpit or fuel system could destroy it. More often than not, the fly-by of a -262 did little more than scaring everyone. The aircraft was no more effective than Germany's Bf 109 and Fw 190.
>
> From mid–1944 to the end of the war, the sky over western Europe and Germany was full of Allied airplanes. The 8th Air Force sent raids of 1000+ bombers plus escorts; the 9th Air Force, comparable numbers of medium bombers and fighter-bombers. The RAF sent out a thousand bombers several nights a week. Credible sources estimate the number of operational jets available to the Luftwaffe to be fewer than 300. The highest number of Me 262 sorties for a single day was 55.
>
> The singular advantage of the Me 262 was that it did not require the high octane fuel used by reciprocating engines. The 8th Air Force effectively bombed German oil refineries, and the shortage of high octane fuel effectively grounded the other fighters.
>
> The jet aircraft brought speed, but the tactics and equipment to exploit speed—radar, computing gunsights, long range precision armament, command and control—were not yet available.

On the side of the Allies, jet development had finally begun. In Britain, the Whittle turbojet engine was in development. The British Gloster E.28/39 single-engine experimental jet aircraft flew on 15 May 1941. German spies were able to provide their government authorities with the exact date and details of the flight.

The United States did not have a program for development of jet engines until the summer of 1941 when U.S. General of the Army Henry H. "Hap" Arnold became aware of the Whittle turbojet engine and its aircraft. The United States had not yet become formally involved in the European conflict, but Gen. Arnold knew he would be responsible for Air Corps resources if the United States entered the war. He recognized the enormous potential of jet propulsion. In late September 1941, following an agreement signed between the U.S. Secretary of War and the British Air Commission's Sir Henry Tizard, one Whittle W.1X engine and drawings of the W2B were shipped to the United States under absolute secrecy. Whittle came to the United States for a few months and resided under an alias, "Frank Whittley," to supervise early work.

Gen. Arnold directed that an engine developed from the Whittle engine be built by the General Electric Company in Lynn, Massachusetts. The first turbojet engine in Germany was not produced by an aircraft engine manufacturer, but instead by airframe builder Heinkel, and Arnold did not choose an engine manufacturer for the American turbojet engine. He selected General Electric because it was a company with great experience with turbo-superchargers, a concept that had been used on many aircraft since 1918 after a first demonstration by Sanford Moss. At the same time that the General Electric engine was being developed, a twin-jet airframe was constructed by Bell Aircraft Company in Buffalo, New York.

A few months later, on 18 March 1942, the General Electric-built Whittle engine GE I-A ran on the test stand. On 3 October 1942, just a year after the turbojet engine development began in the United States, the Bell Airacomet (XP-59A) equipped with two GE I-A engines began test flights at Muroc Dry Lake, California (later, Edwards Air Force Base) and reached flight altitudes up to 10,000 ft. The United States finally entered the Jet Age.

The first engine on test stand in the Bartels and Becker garage, Göttingen.

Max Hahn with the first engine.

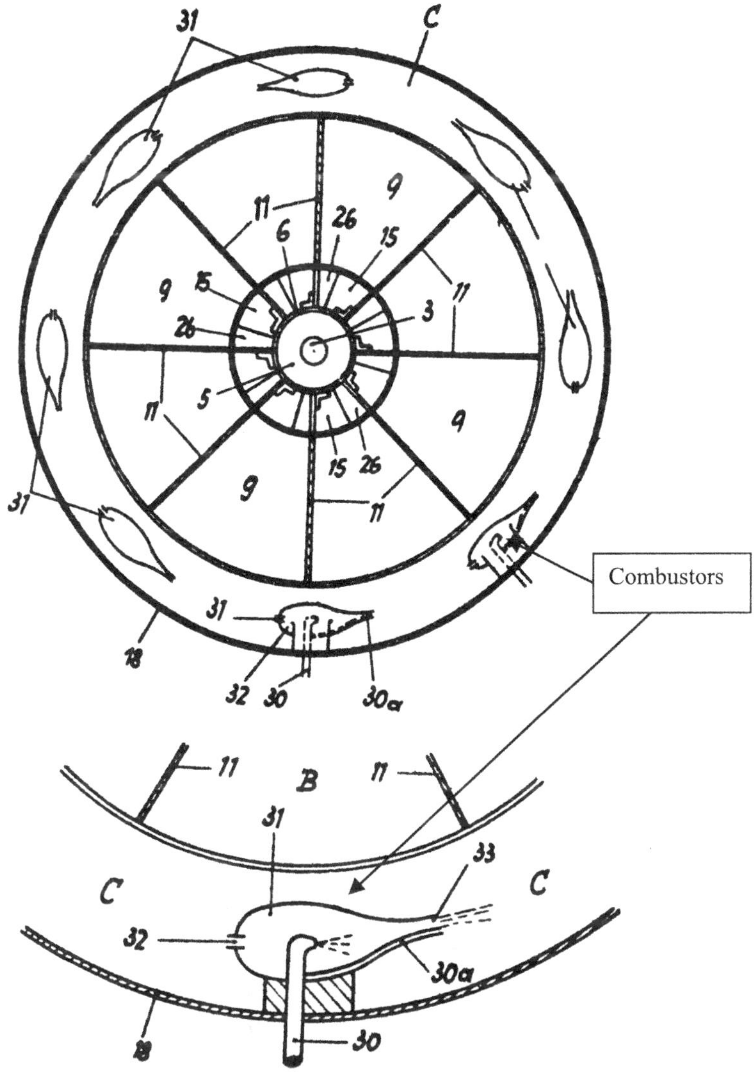

Hans J. Pabst von Ohain, turbojet patent 317/38, 10 November 1935. (Arrangement, Erik Prisell, February 2000.)

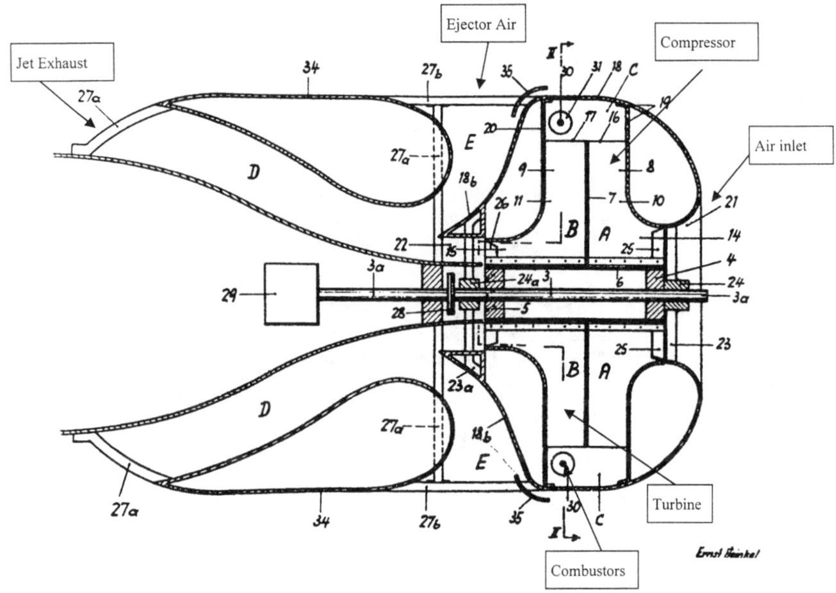

Hans J. Pabst von Ohain, turbojet patent 317/38, 10 November 1935. (Arrangement, Erik Prisell, February 2000.)

Logo for Heinkel Flugzeugwerke, Ernst Heinkel's company, with flying wings and waves, indicating production of seaplanes.

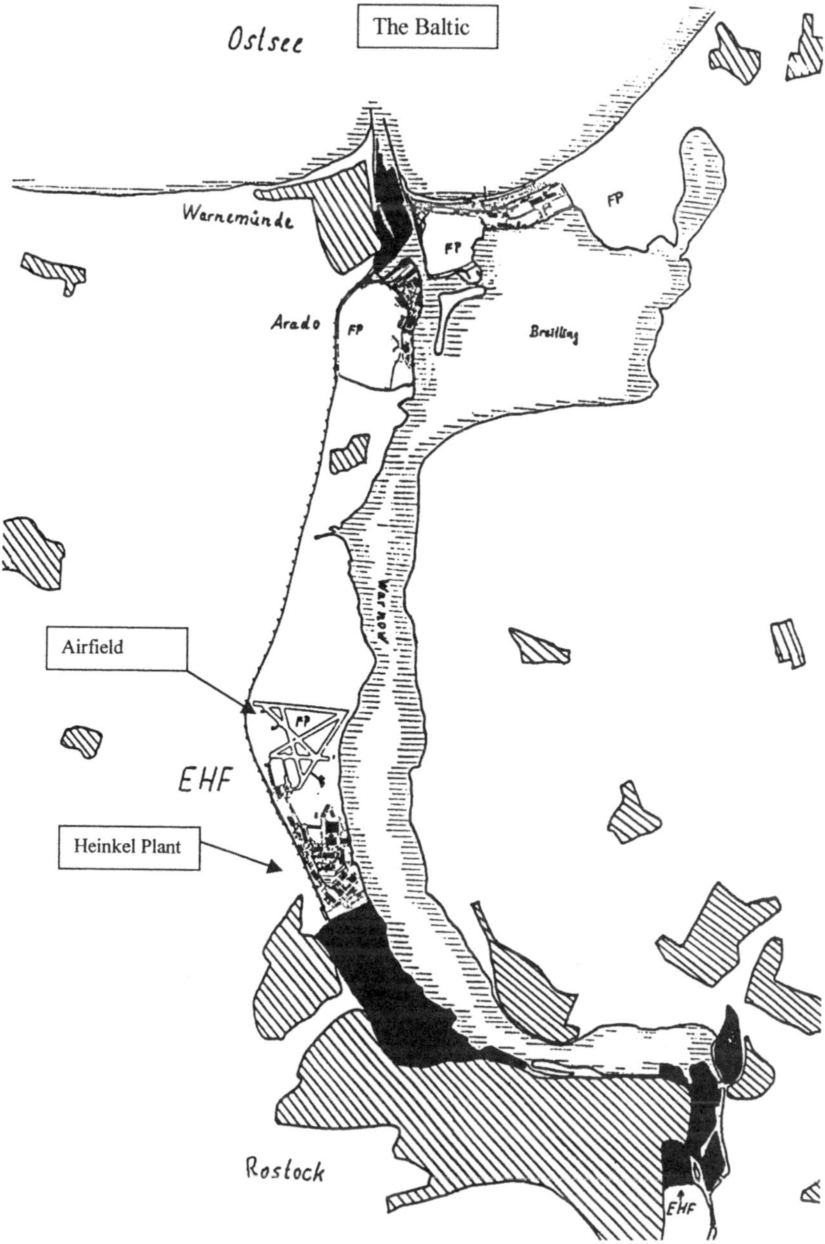

Map showing location of Heinkel factory complex and airfield on the Warnow River between Warnemunde and Rostock. Note the location of the Arado plant.

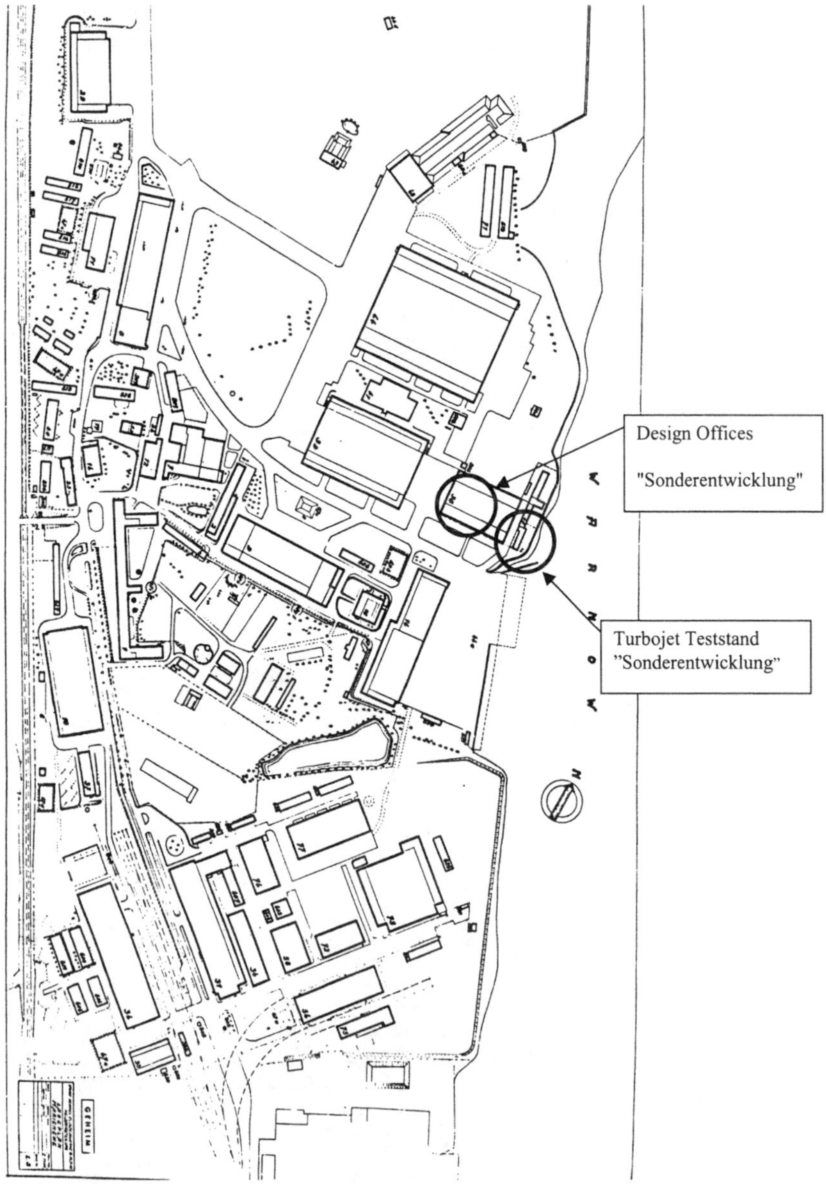

Map of Heinkel factory complex near Rostock, with location of experimental jet engine design offices and test stand on the Warnow River. (Source: Archive of Volker Koos; Description: Erik Prisell.)

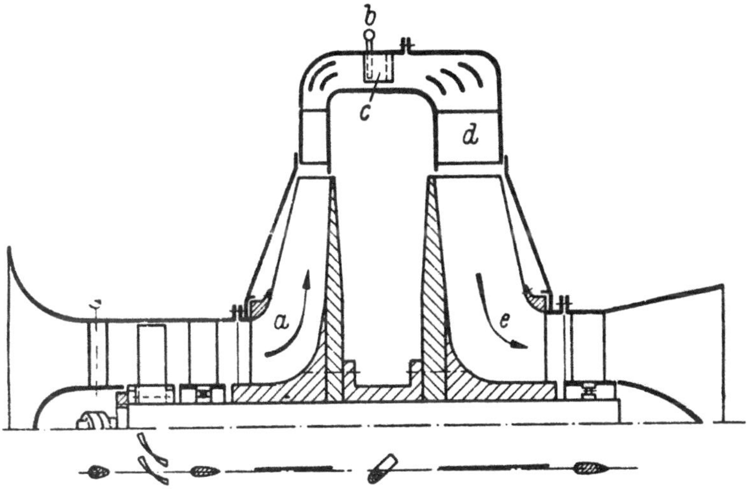

Radial turbojet (He S-1) with hydrogen (built in 1936, tested in March 1937). Radius of rotor=1 ft, thrust=250 lbs, 10,000 rpm.

He 178 aircraft built for HeS3B engine. First turbojet flight, 27 August 1939.

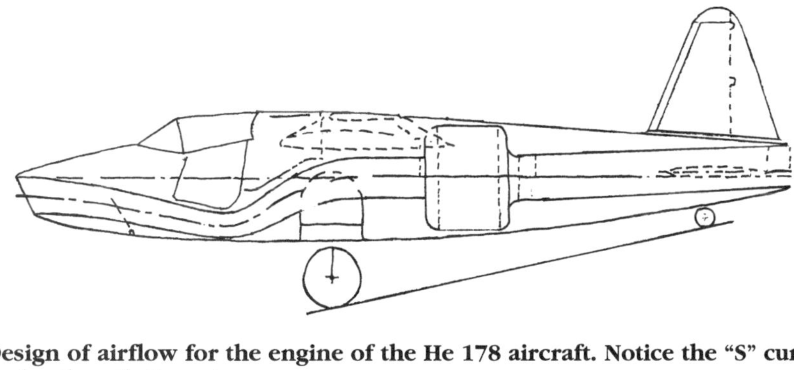

Design of airflow for the engine of the He 178 aircraft. Notice the "S" curve under the pilot's seat.

Ernst Heinkel and Hans von Ohain, 27 August 1939, with a toast to turbojet propulsion.

Cartoon drawing by Max Bentele for Hans's birthday in 1979, imagining Frank Whittle's reaction if he had seen the first flight.

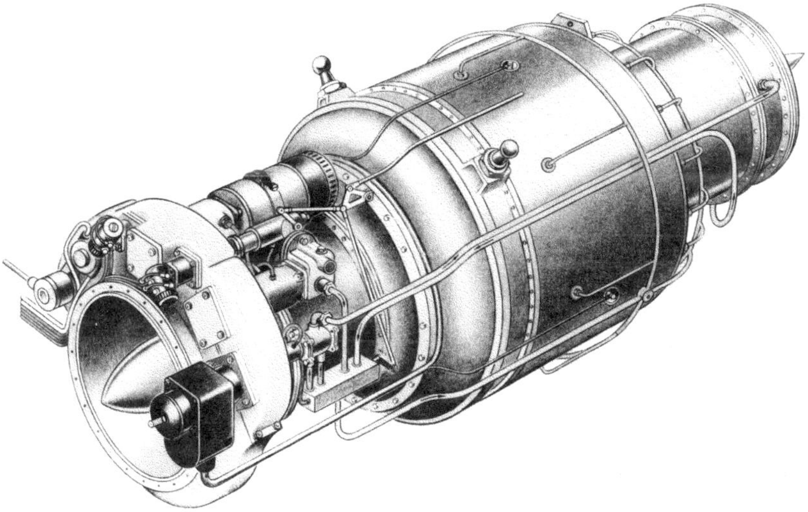

External view of the He S 8A engine showing the installation of various components.

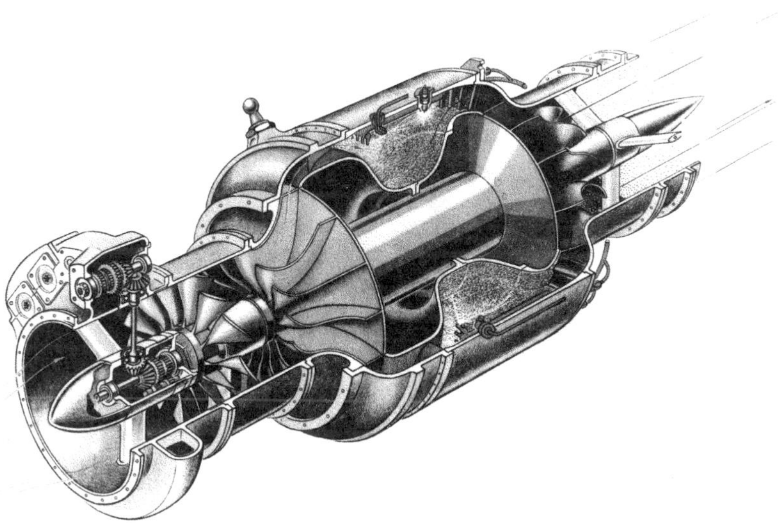

Internal layout of He S 8A engine showing accessory drive, flow inducer, compressor, and radial turbine.

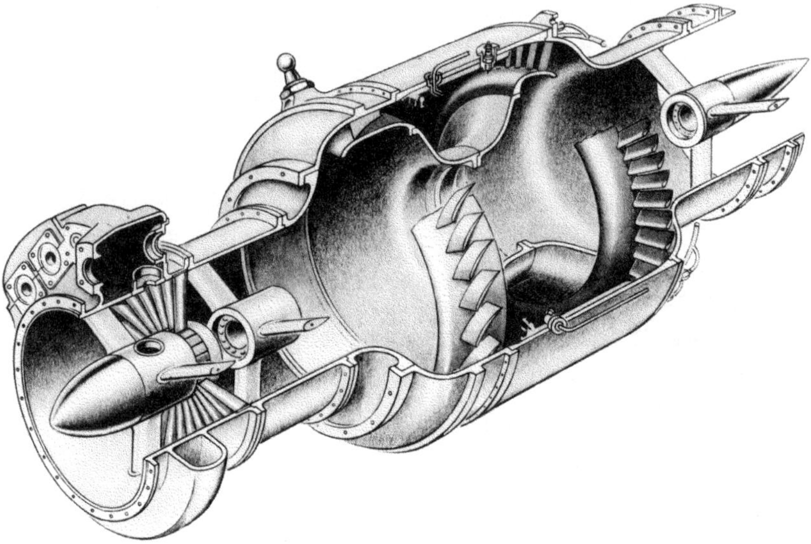

Layout of stationary cascades associated with the flow inducer, compressor, and turbine. (Von Ohain called the flow inducer an "impeller" in the He S 3B.)

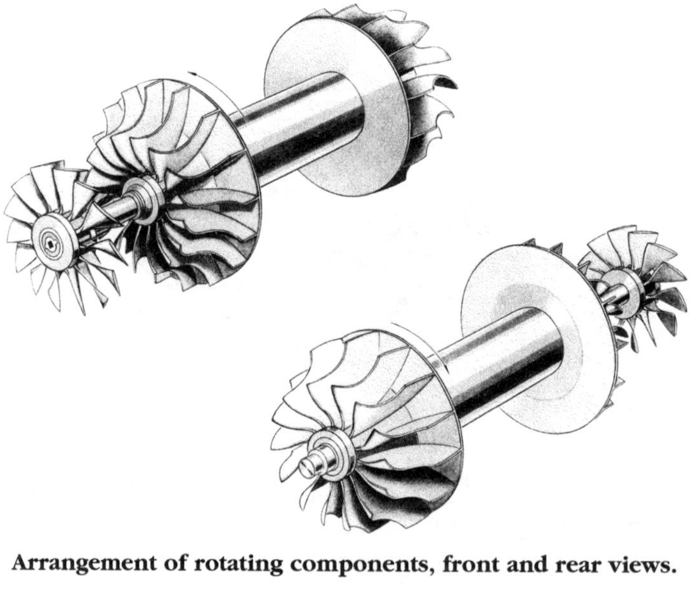

Arrangement of rotating components, front and rear views.

Artist's drawing of the He 280 aircraft powered by the He S 8 engine.

He 280 aircraft ready for first flight. The He S 8A engine's cowling is removed because of leaking fuel.

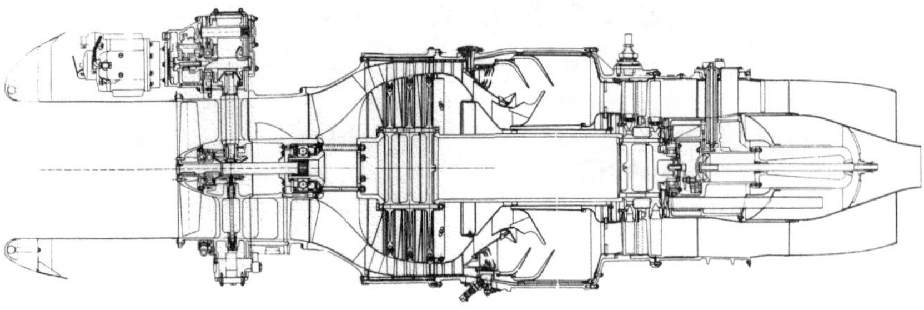

Internal arrangment of He S 011 components.

He S 011 engine.

Chapter 11

Last German efforts and defeat

In 1942, the German Air Ministry wanted a second-generation turbojet engine to be the future main service engine for the *Luftwaffe.* It would be designed to equip all the airframes that would exist after 1945, including the Arado Ar 234, Junkers Ju 287, Messerschmitt Me 262, and some other advanced Messerschmitt versions. It would replace the Junkers 004 and BMW 003 engines. In late 1942, Heinkel-Hirth received the contract for development of that engine, designated the He S 011. It was some consolation to Heinkel for his loss of other turbojet engine projects. The Air Ministry saw the strength of the Heinkel-Hirth engine staff: Hans von Ohain in charge of design and development, Curt Schif for planning for mass production, Max Bentele for components, and the Hirth engineering team and the team that had developed the He S 30 engine. The He 280 V 7 and V 8 airframes were available to use for testing.

The Air Ministry gave top importance to the development of the 011 engine, and its movement to production status. Within a few months after the contract, a conference was called with designers from Heinkel and specialists from the Aerodynamics Research Institute at Göttingen (AVA), who made recommendations for the design of the engine and work-arounds for expensive or scarce raw materials. Helmut Schelp had been appointed a successor for Hans Mauch in the Air Ministry. Schelp believed that the future of turbojet propulsion lay in high pressure ratios. While this improvement could reduce fuel consumption, such an engine was hard to start and accelerate. These difficulties were to be reduced with a new design for which Schelp worked with von Ohain and his team extensively, and contributed to the overall design with some excellent technical inputs for the advanced diagonal [mixed flow] compressor stage, the two-stage air-cooled turbine, and the variable exhaust nozzle system. Ernst Heinkel agreed to follow Schelp's suggestions completely.

Von Ohain wrote:

> The He S 011 developed under my direction had a thrust of 2,860 pounds and a growth potential of 3,500 pounds. The engine achieved pressure ratios of 5:1 of the total compressor, and a growth potential to 8:1 was anticipated. The engine's advanced features included a 50% reaction axial compressor similar to that designed earlier by Rudolf Friedrich for the He S 30, except that for the first compressor stage the 011 used a novel high mass flow, high pressure ratio "inducer [compressor] stage" that had a broad characteristic, now known as mixed flow compressor. We had only five stages altogether for pressure ratio five. This inducer [compressor] stage had been developed in 1941 by Dr. Vanicek and myself. The He S 011 had a two-stage air-cooled axial-flow turbine. The blades were hollow with inserts that directed the internal cooling air and damped the blades to avoid vibrations. This very successful concept was invented by Dr. Max Bentele.

He continued:

> [We did not use] strategic materials such as nickel, which called for a completely air-cooled engine. We protected the blades against oxidation by aluminizing the surface. Nevertheless, the total turbine inlet temperature was at that time (I say it off the cuff) numbers like 1,300 to 1,400 F (700 to 750 C). [For reference, iron turns red at 1000 F and white at around 1800 F.]

The diagonal stage compressor, a cross between an axial and a radial stage, made the engine very difficult to manufacture, but the engine met all specifications 22 months after the contract was signed. Engines V1–V5 were built in 1943, followed by five more in 1944. Nine experimental engines were bench tested, and one or two flights had been made on a Ju 88 test bed, but no engine was flown under its own power. Von Ohain said, "By the end of 1944 and early 1945, all the performance data had been met or exceeded." Production drawings had largely been completed and large-scale production of the He S 011 "A-O version" was projected by the Air Ministry to begin in mid-1945. The end of the war came before the Germans could install the He S 011 in an airframe.

At the end of World War II in 1945, von Ohain's He S 011 engine was one of the most powerful turbojet engines in the world. In competition were the GE I-40, which ran on the test stand in February 1944 with 4000 lbs thrust per front area, and the Rolls-Royce Nene running at 5000 lbs thrust by late 1944.

Aircraft engine development was Hans von Ohain's full-time employment, but he could not ignore what was happening in his country.

At the beginning of the war, the German people cheered at torchlight parades of massed armed forces in the large cities and for news reports of marches into country after country across the European continent. Adolf Hitler was their hero. Victory followed victory. In September 1939, the powerful German *Luftwaffe* had initiated *Blitzkrieg* from the air, decimated the Polish air force, and defeated Poland. In April 1940, Denmark and Norway fell. In May 1940, Holland, Luxembourg, and Belgium were invaded. In June 1940, France fell. On 7 September 1940, Germany sent 300 bombers with 600 fighter escorts to bomb London, continuing with 85 nights of raids. A second front was initiated on 22 June 1941, when Russia was attacked with 170 divisions over a thousand miles of Russian frontier with the help of 2770 *Luftwaffe* planes. Eight thousand Soviet planes had been destroyed by the end of 1941. German tactical bombing, using dive bombers and medium bombers to support troops moving into an area, changed the face of war from the previous centuries of land assaults.

A slow realization followed that war on two fronts was difficult to sustain. German troops met their first failures in Russia, enduring battles with large armies as well as the daunting Russian winter. Germany had unbearable losses in skilled pilots. From December 1942 through January 1943, 500 German aircraft were lost in Russia during attempts to supply troops. German bombers could not reach the aircraft factories in the vast distances of the Russian homeland, and the Russians were able to replace their planes.

As the war progressed, German production did not have the resources to keep up with the overwhelming strength in numbers of the Allied air forces throughout Europe. Germany had no heavy bomber equivalent to the American B-17 and B-24 and British Lancaster, no multi-role fighter-bomber such as the American P-47 or British Typhoon, and no night fighter such the P-61. Shortages of materials, particularly strategic metals, became a manufacturing problem. Delay of the development of Anselm Franz's engine for the Me 262 aircraft was caused by the need to produce with only available materials. The weak leadership of the Air Ministry failed to provide the needed strategy for production of aircraft for the German war effort, and its staff lacked necessary qualifications to set aircraft development requirements. Virtually none of the development programs initiated by German design and development firms after 1939 produced an effective aircraft.

The *Luftwaffe* had no plan for strategic bombing, while Allied bombing of factories, bridges, and transportation in Germany became devastating. In April 1942 Heinkel's factories at Rostock and Marienehe were heavily bombed. At this time, von Ohain had already gone to the Heinkel-Hirth factory at Zuffenhausen. The picturesque, prosperous towns and farms of the 1920s that had been changed into a factory complex from north of Rostock to Warnemunde, an "industrial monomania" for the production of aircraft, were turned into rubble. Heinkel was forced to move his testing facilities from Rostock to Schwechat, near Vienna. In August 1943 at Regensburg every building of the Messerschmitt plant was damaged by bombings. Remarkably, there was rapid recovery of production capability at that location, made possible by the German government's policy established early, to scatter factory buildings in campus style rather than placing all work in one building.

In an unrealistic philosophy for wartime, the Air Ministry encouraged respected scientists in many industries, including the aircraft industry, to initiate a wide variety of technological concepts. They generously financed projects without providing careful management or coordination of efforts. German people whispered rumors about "super weapons" being developed in laboratories. Magnetic mines for the Navy! Radio controlled rockets! A gas that could be let loose in an area to destroy all life in seconds! A fantastic "atomic bomb!" Death rays! A tunnel being built secretly under the English Channel through which the German troops would march to Britain! Some of these developments were horribly real, as the Allies learned after the war ended; the tunnel was only a wish. Descriptions of the variety of proposed German secret weapons make for fascinating reading and in later years were raw material for sensationalized science-fiction writing. "Evil genius" German scientists were actually imaginative scholars given freedom to create, with funding from the government. Hans von Ohain was one such scientist working in one location in the jet engine program, which was only a small part of the total *Luftwaffe* effort.

Von Ohain was part of the industry that produced 40,000 aircraft for Germany during the war, 60 percent of which were fighter planes (few were jet-propelled). Jet aircraft were notoriously short-ranged, but showed potential effectiveness as fighter, bomber, and photo reconnaissance. Short-sighted German authorities discounted the potential for this new weapon. Even after the turbojet-powered Messerschmitt Me 262 and the Arado Ar 234 aircraft were developed, no priority was placed on their production to field them earlier and in sufficient number.

Albert Speer had been appointed in 1942 to be Hitler's minister for armaments. By 1944, he viewed defense as the first priority for Germany, instead of the national policy of aggression. Speer's minister for air equipment was Karl-Otto Saur.

Saur conceived the idea of using the dramatic new form of propulsion, jet power, to power a new design, the *Volksjäger* (people's fighter). The turbojet aircraft already in production required extensive resources and time to produce. He envisioned a simple single-engined jet fighter that could be mass produced even in the shambles of the German aircraft industry by semi-skilled and unskilled labor, using readily available materials, and put into service quickly. Boys from the Hitler Youth, the *Volkssturm,* organized in the last year of the war, would be given rudimentary flight training on gliders and be sent to fly this new turbojet. No maintenance plan would be needed for the planes because they would be built in vast numbers and used and discarded, rather than repaired. Such a plan would have been laughed out of existence in less desperate times.

An "Emergency Fighter Competition" was prepared. Specifications were drawn up by the Air Ministry and issued to Arado, Blohm and Voss, Fieseler, Focke-Wulf, Heinkel, Junkers, and Messerschmitt. They were as follows:

1) Maximum speed at low and medium altitudes with minimum speed of 466 mph at sea level.

2) Unassisted takeoff within less than 1970 ft.

3) Minimum production build-up to obtain large-scale production in the shortest possible time.

4) Extensive employment of wood for the structure.

5) Production of the type not to interfere in any way with the manufacture of the Me 262 and Ar 234.

6) Exploitation of dormant production capacity made available by cancellation of other production programs.

7) Only one turbojet to be employed.

8) Armament to comprise two 10-mm Mauser MG 151 cannon interchangeable with two 30-mm MK 108 cannon.

9) Radio equipment to comprise FuG 24 R/T and FuG 25a IFF.

10) Protective armor against only frontal attacks.

Heinkel had an effective network of informers who kept him in close touch with Air Ministry discussions. He obviously had begun the design of the He 162 early, when he first heard that there were prospects for such a plane. The chronicle of Heinkel's fielding of the He 162 is remarkable to anyone in the field of acquisition. The time of record was only five months from the date that the aircraft specifications were issued to the date of the first production delivery:

On 8 September 1944 the Air Ministry issued the specifications. For weeks many personnel in Heinkel's design office did not leave their drawing boards, day or night.

On 20 September, 12 days later, the Ernst Heinkel Company had demonstrated a mock-up plane. Heinkel could fulfill the list of specifications, particularly the criteria for a large "dormant production capacity," and Heinkel was awarded the contract.

On 24 September construction of prototypes began with detailed drawings being produced at the same time.

On 6 December the first prototype was flown at the plant at Schwechat.

On 6 February 1945 the first He 162s were delivered to an operational *Luftwaffe* unit.

The airframe was tiny compared with the Me 262 and Ar 234 and was largely constructed of wood. It was produced at the underground salt mine at Hinterbruhl. Heinkel first considered using the von Ohain He S 011 engine to power the new plane but found that it was too large and heavy. The smaller, lighter BMW 003 engine was selected for the prototype and first production models. The von Ohain He S 011 was designated for the He 162 aircraft in advanced versions.

The He 162 *Volksjäger* was also called the "Salamander," so called for the mythical ability of the salamander to withstand fire. It flew in limited service on 4 May 1945. Heinkel's remarkable effort to get the plane into production was futile.

After the war British pilot Capt. Eric Brown flew the He 162 at Farnborough. His description of the plane is found in the June 1972 issue of *Air Enthusiast* in an article called "Mastering Heinkel's Minimus." He gives this opinion:

> It was like all German jets, a superb aeroplane in its element but quite a handful to take-off and land. It would have been an effective gun platform, and its small dimensions would have

> made it difficult to hit. Even if somewhat underpowered, it had a good performance. It would have run rings around the contemporary [British] Meteor. It was [however] no airplane to let embryo pilots loose on, and it would have demanded more than simply a *good* pilot to operate it out of a small airfield. Nevertheless, as a back-up for the formidable Me 262, it could conceivably have helped the *Luftwaffe* to regain air superiority over Germany had it appeared on the scene sooner.

Close reading of his review shows conditional statements and qualifiers in an attempt to put the He 162 in the best possible light. Most accounts describe it as a "killer" machine. The prototype disintegrated in an early demonstration flight, and one of Capt. Brown's colleagues at Farnborough died in a postwar evaluation flight.

In the spring of 1944 German factories were set up, camouflaged or hidden in caves and abandoned mines, with their goal to produce between 2000 and 3000 jet engines per month. More than 1400 Me 262s were delivered to the *Luftwaffe* between March 1944 and 20 April 1945, but only about 300 got into operation. The rest were destroyed by Allied bombings.

Hitler continued to insist that his jet plane, the Me 262, should be used as tactical support for ground troops as a low-altitude bomber. This was not unreasonable because bombs were relatively small, and mechanisms for carriage and release were not complicated. However, Hitler discounted its superior possibility as a high-altitude fighter and interceptor. Gen. Adolf Galland, with his influence from being an "ace" in the war, tried to convince Hitler that the plane as a fighter was essential for air defense. Near the end of 1944 a small group of *Luftwaffe* wing commanders staged a heated protest against Göring for his withholding of jet aircraft to act as interceptors. In response, Galland was put in charge of a squadron made up of the cream of fighter pilots from all of the *Luftwaffe* to fly the Me 262s in a last glorious effort against the vast armadas of Allied bombers. Only in the last months of the war were the Me 262s used as fighters. Me 262s destroyed an entire flight of 12 American bombers in January 1945. One estimate states that the Allies lost 300 bombers to jet aircraft during World War II. The number is inconsequential compared with the number of aircraft launched against Germany by the Allied air forces.

By late 1944, it was apparent to all of Germany that Hitler's war was hopeless and could not last very much longer. Widespread Allied bombing of production facilities had caused extensive damage to fuel resources and rail lines, as well as military targets and population centers. The troops were demoralized, and the German citizens were suffering without food and shelter. They secretly listened to Allied radio broadcasts and had mixed emotions, grief for their country and relief that the Nazis might someday be defeated. Before the end of 1944 transportation was paralyzed. Telephone lines were not working. Most store shelves were empty. As they drank their chemical beer, the Germans spent time retelling old jokes about the regime and then swallowed the latest propaganda, hoping against reality.

In March 1945, the Allies were invading Germany from all directions. Heinkel closed most of his factories. Von Ohain and his team in Zuffenhausen near Stuttgart continued with design and development of jet aircraft power units and components until Hitler published an edict:

> Since the German people have been too weak to attain victory, Germany must be destroyed. Nothing must be left for the enemies. Everything must be blown up or burned: electrical power plants, gas and waterworks, factories and warehouses, bridges and buildings of any size, dams and waterway locks, railroads and rolling stock, all vehicles, all farm animals, and supplies of any kind. Even the autobahns shall be destroyed.

Albert Speer, Minister for Armaments, tried to reason with Hitler that the German people needed to have some resources to live on after the war ended, but Hitler, in his final crazed condition, would not listen to him. Speer took the matter into his own hands and recommended that the edict should not be followed too literally.

There were nine He S 011 engines left in Heinkel's Stuttgart plant in the process of bench testing. When Allied forces drew near, the von Ohain team acted in compliance with Hitler's order. They covered the engines with magnesium and very effectively set them ablaze. Von Ohain was later interviewed about his feelings when seeing the results of his hard labor going up in flames. "May I say here something very bad?" Von Ohain chuckled. "We enjoyed the destruction because we thought that when the Americans came over we would remanufacture them. We would have a transition period where we could review the results and the consequences. You see our morale...." And he laughed. They had worked on the engines for a long time. Changes that needed to be made could be more easily implemented by building a second time.

When asked if he was glad Stuttgart was in the American zone, not the Russian one, von Ohain replied, "Yes, oh, my, yes. It's a funny thing. The [Nazi] Party had a hard time to inject into the population what you may call a little bit of hate against the Americans. The Germans were very happy to get the Americans. It wasn't in comparison to the Russians. They liked the Americans, period. Göbbels? They didn't believe him any longer. The population knew the truth."

After the destruction of the 011 engines at Stuttgart, Heinkel directed von Ohain and his team to go to the small village of Kolbermohr between Bad Aibling and Rosenheim because of Russian troop advances. The plant was an abandoned cotton spinning factory *(Baumwollspinnerei)*, and was already being used as a dispersal area for some BMW work from Munich. The Heinkel team selected test sites in the area and accepted delivery of some materials. Construction of facilities had only started before the area was taken over by the American forces.

Von Ohain describes the Allied occupation of the area:

> This was shortly before the official surrender [of this area] in April, during the beautiful days in March. We were overrun by the Americans in the little Bavarian village-city of Kolbermohr. It was on a beautiful little creek, and you saw the Alps silhouetted against the sky. It was absolutely a dream of beauty. We saw the P-38 twin-fuselage Lightnings flying over. That little village didn't attract their attention. We were sent there, relocated to be away from any attacks. We were to build a field test facility for the testing of 011 engines, something that we felt was not really realistic. One had the feeling, "It is all over." There is an instinct in you that says, "There's no future in building a research facility for these engines." It was a very tough thing, because if you said it too loud or to the wrong people, it could have terminated your life instantly. They [the Americans] came through and searched us for knives and guns. I had a gun, but I didn't tell them that. I buried it later on.

Just behind the occupation forces a team appeared that seemed to be ordinary Army officers who were, in fact, intelligence experts and highly skilled civilian scientists trained to identify and seek out certain talented German scientists. When the Allied authorities moved into the area they looked for engine development records, technical data, and some key personnel.

The Allies were quick to find von Ohain in the little village and brought him back to Stuttgart for questioning. They found it was very effective to interview scientists on site at the factories and laboratories where they had been working. Von Ohain said Cliff Simpson, a captain in the Army Air Corps, was the first American to interview him. (Von Ohain met Simpson again at Wright Field shortly after he arrived in 1947.) Simpson had no apparent reaction to the fact that von Ohain was the creator of the first flight of a turbojet engine. Von Ohain shrugged, "He was interested when the first turbojet flight was, but mildly. He was more interested in the 011."

Von Ohain said, "I pointed out the performance characteristics of the 011 engine. They wanted to know what the source was, what the efficiency, the performance data and where we stood in the production. They were very intrigued by the blades, how we had built them very cheaply, and by the damping of these hollow blades made with no strategic metals."

In 1982, Cliff Simpson recalled their meeting:

> I was a Captain the U.S. Army running wildly around Germany as a member of ATI [Air Technical Intelligence] hitting intelligence targets, messing with ADIK [?], 'T' Force, Special Services and such. I first met von Ohain in Zuffenhausen. His English was better than my German, which is not a compliment. There were two people trying to help us communicate, a Heinkel employee named Wolfe and my jeep driver, Liepold. Hans and I did communicate, but I must say the conversation was less than flowing.
>
> Hans was working on getting the 011 engine into production. Even today a tour through this design with its designer is a real treat. Many novel features, many problems that a little nickel, chrome, and cobalt could cure. I can vouch for its running, having tested it in the BMW altitude facility at Munich. The data was shared by the U.S. and U.K., and the parts to build the test hardware were collected from all over Germany by thirty-seven of us chasing them down in the August-September 1945 time frame.
>
> Both Hans and I tried very hard to communicate and did succeed in building trust and respect in each other. The doubts that existed were based on my rather shallow knowledge of the turbine engine which was based on my memory of 'Stodola' and Sir Geoffrey Smith's little book purchased at a street bookstall in

> London. [A. Stodola wrote the classic book, *Steam and Gas Turbines,* in 1927, with an appendix on gas turbines and the future of heat engines. G. Geoffrey Smith, M.B.E., wrote *Gas Turbines and Jet Propulsion for Aircraft,* first published in about 1945. It is considered essential reading for the basics of jet propulsion.] I was not sure exactly what Hans had done, but as a result of our conversation in garbled German and broken English I concluded that one of the things he had not done was stand around twiddling his thumbs.

The discussions between them took only a day and Simpson left, but before he went he took the time to tell von Ohain about jet propulsion as it was being studied at Wright Field. Von Ohain said:

> He was kind enough to tell us a little bit about the work at Wright [Field]. We were happy to hear that there was some work going on in gas turbines for aircraft. He told me of rpm values which were very high, and I said "Well!" But it was in the realm of possibility. It must have been the Whittle engine which was being made by General Electric, the GE-1. I found out later that it was a W-2 Whittle engine. It was reproduced and tested at Wright Field. We were thinking, "Well, that's hard to believe," but it was correct.

Russians moved into the area near the Baltic Sea in early May. They seized and dismantled what was left of the Heinkel works at Marienehe, carrying off the assets as war reparations, and they blew up the remains of the bombed Heinkel factory at Rostock. Nothing remained of the extensive Heinkel production factories. Today a single stone commemorates the historic first flight of the He 178 aircraft. A plaque on the stone says:

27 August 1939
Start des ersten Strahlflugzeugs der Welt,
der He 178, entwickeit in den
Ernst-Heinkel-Flugzeugwerken Rostock
mit dem Hans v. Ohain Triebwerk He S 3b

(Translation: 27 August 1939, Flight of the World's First Jet Plane, the He 178, developed in the Ernst Heinkel Aircraft Works, Rostock, with the Hans von Ohain He S 3b Engine.)

Daily, weather permitting, the American 8th, 9th, and 15th Air Force planes bombed factories and railroads in the heart of Germany and Austria. In the last months of the war the air forces concentrated on destroying German oil refineries and synthetic fuel plants. Soon, German aircraft had empty gas tanks.

Hitler committed suicide. On 7 May 1945, the German High Command finally surrendered unconditionally, and, within days, Allied aircraft began dropping food packages and supplies to the areas in Germany where they had bombed just a short time before.

Hans von Ohain was separated from his family during the war, but he was concerned about them. His parents and brother Herbert were in Berlin, a favorite site for Allied bombing. Wolf, his younger brother, was in the army. Hans was deeply upset when he learned Wolf had been sent to Russia. He was grateful that he himself had been exempted from military duty because he was considered to be essential to the war effort, a scientist employed in an industry valuable for the building of weapons to strengthen the new Germany.

Hans von Ohain's father, Wolf von Ohain, had been a very successful businessman during the 1930s. In an interview, Wolf Jr. explained that his father had contracts from theaters and prisons and jails, to "all government," for light bulbs. Of his father he said, "He must have been extremely charming because he made a lot of money." He had to give up his business because by the late 1930s there was pressure from the Nazis to control all parts of the economy by excluding the "middle man."

Wolf Sr. retired and was living comfortably. Unexpectedly, during the war he was informed that he was to be reactivated into the military. His military record was outstanding and he was strict, straightforward, and honorable. When he asked what an old man such as himself might do in the military, he was told that he would be promoted to colonel and become commandant of a prison camp. This was an order that he could not refuse, but he did not want the job. A doctor who was a friend wrote a certificate that Wolf's arthritis was severe, much more severe than it actually was, and he would be unable to serve. The authorities accepted the statement and Wolf's reinstated military career ended before it began.

Hans's parents continued to live in their house in Dahlem during the war. Storage rooms in the lower level of their house, where the kitchen was located, were their bomb shelter. Fires on upper floors of houses were frequently deadly for those hiding in basement bunkers because the fire depleted the oxygen in the house and smothered those hiding in the basement. Germans had experience with handling incendiary and phosphorus bombs and knew that it took a while for these weapons to explode. During one attack, an incendiary bomb crashed through the tile roof of the von Ohain house. The elderly Wolf hurried upstairs and threw it out of the window, saving the house. Some months later the von Ohain home was slightly damaged when an air raid destroyed the house next door.

Hans von Ohain's older brother, Herbert, was a corporate attorney in Berlin during the war. His position and age kept him from being summoned into military service. In the last year of the war, when there were no longer young men left to draft, Herbert was conscripted into the local defense group, which was something like the National Guard in the United States.

Wolf Jr., Hans von Ohain's younger brother, had been considered a gifted student. He studied in Britain and France during his teens and became fluent in English and French. He looked forward to studying for a career in architecture. As most young men in Germany in 1941, he had to lay aside his future plans. Upon completing studies at *gymnasium* he was immediately drafted into the artillery. His assignment after training was the Russian front in an advance force just outside Moscow. He said that these troops were very disciplined and that the fighting was very fierce. He saw horrible acts where whole villages were burned out with artillery fire. As for equipment, he reported that the German army had massive tanks and heavy trucks that were quickly mired down in the knee-deep mud when the fall rains came. Roads in Russia did not compare with those in Germany.

In an interview in 1999, Wolf sighed, "I remember the mud in the fall, and the terrible winter that followed." Wolf told stories of the cold. "We were poorly equipped and had no cold-weather clothing. When we found a dead Russian we buried him, but took his felt boots that had a broad base and did not sink in the snow." They did not take the Russians' coats because they were most often infested with lice.

While in Russia, in the same location where his great-great grandfather had been killed fighting for Napoleon, Wolf was wounded. He was wounded two more times during the war. Each time he recovered in Germany before being sent back to his unit. In 1945 he was captured by the Russians and interned in a prison camp in Czechoslovakia. He managed to escape, traveling by night and hiding during the day. With the aid of a compass he walked back to Germany and surrendered at an American prisoner-of-war camp.

Chapter 12

Paperclip

Years before the end of World War II, the United States began planning actions that would follow Germany's defeat. Financial reparations, as had been paid after World War I, could not be extracted from a country that had suffered wholesale bombing and devastation. All of the Allies decided that the most valuable "spoils of war" from defeated Germany would be knowledge about German aircraft and equipment. Even better than gathering paperwork and armaments, which could become obsolete, would be acquiring the scientists who had designed and built German armaments. Before the final capitulation of Germany in May 1945, the Allies—the United States, Britain, France, and Russia—were in fierce competition to locate and evaluate scientists who had made significant contributions to the German war effort. As an area was liberated, the goal was to find and move the scientists out of Germany to the home countries.

Gen. Dwight D. Eisenhower, commander of U.S. forces in the European theater until 1945, believed that it was the role of the American military to take the initiative in the recovery of these individuals. Unfortunately, the military did not immediately know the names of the most valuable German scientists. Vannevar Bush of the U.S. Office of Scientific Research and Development issued an angry statement to the military leaders, saying, "In scientific circles, the names were as familiar as Churchill, Roosevelt, and Stalin are in political circles." He offered the information available at his agency.

The U.S. government had no precedent or policy for deriving intellectual reparations following a war, and there was no clear term to describe the purpose. "Exploitation" was used and later rejected as too blunt. "Denial," that is, getting a scientist to the United States in order to keep him from going to the Soviet Union, was used frequently, to "prevent, check, restrain, confine, and contain the military power of the Soviet Union." The government had no structured plan to bring selected scientists into the country in spite of the urgent need to remove them from Germany before the Soviets might kidnap or lure them away. They knew there was no time for lengthy paperwork, the formal waiting period required by legal immigration regulations, or the limited quota of Germans authorized for immigration each year. A decision was made that the scientists would be secretly "smuggled" into the United States without visas under military custody, circumventing the normal procedures. If the scientists decided to remain in the United States, some kind of formal entry into the country would be worked out at a later time.

As part of the preparations for the invasion of Europe, an "Exploitation Division" of the Air Technical Intelligence Command had been established. The U.S. Army Air Corps (USAAC) set up an intelligence school at Harrisburg, Pennsylvania, and a technical intelligence school at Wright Field in Dayton, Ohio. Teams of military and civilian experts were assembled who had technical knowledge in the disciplines identified. The United States trained these agents to go into Germany to identify scientists, documents, and equipment that would be useful. After training, they were transferred into war zones and uniformed as combat troops. Their first assignments were to follow behind the ground armies and perform on-the-spot examinations of captured equipment, documents, and aircraft. Interesting material was shipped to London for preliminary review and then to the Air Documents Research Center at Wright Field, Dayton, Ohio.

In April 1945, the U.S. Strategic Air Force in Europe was assigned to coordinate exploitation of German scientific advances and technology under the code name "LUSTY" (LUftwaffe Secret TechnologY). (Theodor von Kármán, who traveled to Europe as part of the first LUSTY team looking for scientists, called the code name "unlikely but pleasant.") In July 1945, the U.S. government's Joint Chiefs of Staff sanctioned "Project Overcast," a temporary exploitation of a maximum of 350 "chosen, rare minds" to assist with efforts to defeat Japan and provide some postwar military research. The number included both scientists and technicians. A few Germans were recruited and brought to the United States, but, through inadequate planning, many were not provided with facilities or purposeful work.

The American public was disinterested in the politics of the effort to find and recruit German scientists and recover documentation and equipment. President Truman and members of Congress were slow to acknowledge the need for funding and legislation to bring scientists from Germany. The war with Japan was continuing. Recruiting a few German experts who might provide assistance with advanced armaments to help defeat the Japanese seemed like a good plan, but in the long range, a "lasting peace" was the most important goal for the civilian population. Gaining scientists who were experts in building Germany's vast array of armaments did not meet that goal. President Truman stated that he didn't want "our boys," veterans returning from war, to be displaced in jobs by former enemies.

The early weeks following the end of the war in Germany resulted in exciting tales of plots to seize and carry off files of scientific information for one's own country, sometimes right under the noses of another country's representatives. The intelligence teams were frustrated by the plundering done by their own troops and by resentful former prisoners of war who were the first ones at a scene and destroyed materials. When German scientists were deprived of work they simply disappeared and went home and were difficult to locate.

First priority for the investigators was finding rocket scientists and scientists with atomic and nuclear experience. All the countries had become aware of the advanced technology developed by the Germans—long-range rockets, guided missiles, and the flying bombs used effectively in attacks on Britain. Intelligence sources knew the Germans had very sophisticated laboratories studying atomic power and did not doubt their ability to build a nuclear bomb. The U.S. military wanted, at all costs, to keep the knowledge of the scientists for these projects from falling into Russian hands.

The next highest priority was for the scientists who had led Germany in jet propulsion and aircraft design. After the first flurry in the search for the rocket and nuclear scientists and their emigration from Germany, a short list of the scientists "Most Desired for Exploitation" was published by the United States. This list included Hans von Ohain, jet engine designer; Adolf Busemann, expert in supersonics and shock wave apparatus who had, at the 1935 Volta Conference for aeronautics, proposed the swept-back wing for aircraft design; and Woldemar Voigt, the Messerschmitt aircraft designer.

Russia was acting less the ally and more the competitor. Some scientists living in the Russian zone of Berlin were summarily roused from their beds, their household goods packed into moving vans, and, with their families, taken at once to Russia. Erich Warsitz, the test pilot for the first rocket flight and the first jet engine flight, reported that he was "kidnapped" from Berlin and spent several years in Russia before escaping. In September 1952 Warsitz wrote, "Comment on this chapter of my life is superfluous because you surely know or can imagine what it is like to deal with men who have fought the whims of the wolves of the steppes for years."

Finally, in 1946 President Truman approved a larger program, Project Paperclip. The Paperclip program was named for the action when the paperwork for a candidate accepted by the United States was completed: a paperclip at the top of the file indicated that the scientist was cleared for transfer. Project Paperclip increased the number of scientists who would be offered the opportunity to come to the United States, provided an offer to bring the scientists' families to the United States, and allowed an option for the scientists either to return to Germany after a given period or gain U.S. citizenship. The project's expanded charter acknowledged that the scientists would provide "willing cooperation and maximum contribution" if they were treated as "honored guests," paid a good salary, and had their families at their sides.

The United States needed the German scientists. There were not enough highly trained American scientists to fill the technical jobs. The military draft during World War II had taken America's young men away from their college educations, and it would be years before their educational level could catch up to a useful standard. Patriotic American scientists who had been working in government laboratories during World War II resented that the research laboratories were being run by veteran military officers who had very little technical education, yet were able to make final decisions for highly educated scientists. They left to take more lucrative jobs with industry.

Clarence Lasby's book, *Project Paperclip,* makes this statement:

> The Paperclip scientists, as an immigrant group, were unlike any previous newcomers. Their numbers were small, comprising hardly a ripple when viewed against the sea of historical migration. They were remarkably homogeneous. The places of birth of the 475 who were in the United States in early 1948 were as follows: Germany, 428; Austria, 16; Czechoslovakia, 7; Poland and Russia, 5; Switzerland, Estonia, and Free State of Danzig, 3; Hungary, 2; Belgium, Italy, and Yugoslavia, 1.
>
> They were extraordinary for another reason. "It might not be an exaggeration to say," wrote one of them, "that the Paperclip

> program was the first in American history where an entire group of immigrants were far above average in their intellectual capacity and mental heritage."

Each German scientist was assessed for his potential value in "the national interest." His background was checked concerning participation in Nazi political affairs. "Denazification" was being enforced by American authorities for all Germans, frustrating the efforts of the military officers urgently recruiting valuable scientists to keep them from going to work for the Russians. Most scientists were listed on paper as former members of the Nazi Party. Their activities had to be screened to discern who had been active participants in Party affairs and who had joined "for convenience" or "as required by the job." Some had never been Nazi Party members. Max Bentele was interviewed by military authorities and admitted that he had been a Nazi but had left the organization when he became disillusioned. The officers shouted in triumph, "At last we have a Nazi! Every other person we have asked said that they had never been a member." Interviews with the scientists showed that the scientists were remarkably apolitical. They were more concerned with their future professional status.

Not all recruitment efforts worked out to the advantage of the Americans. Siegfried Günter, former chief aircraft designer at the Heinkel Company and probably the world's greatest expert on aerodynamic design, offered his services to an American officer in 1945 only to be turned down because, as the officer said, "We already have good designers in America." Günter was hired in Russia. Some scientists fled to South America to set up businesses there.

Von Ohain was approached a month after Germany's surrender by an officer from the USAAC, probably under the terms of the earlier Project Overcast. He stated that they had been advised of the value of von Ohain's work and would be providing him a contract to come to Wright Field to continue with propulsion research. Von Ohain was interested, but weeks went by and he heard no more.

There was no work relating to aircraft or engines in Germany in the summer of 1945. Von Ohain set up a small business doing consulting services in air conditioning and refrigeration. Ernst Heinkel cancelled contracts with his staff. This letter was sent to von Ohain:

Dr. Hans-Joachim Pabst von Ohain
Korntal, Leonberg County
18 July 1945

Owing to current conditions, it is not possible to continue development work in your specialty under the terms of your contract of 12 June 1939.

Per your wish, we will dissolve the employment contract between you and ERNST HEINKEL COMPANY effective 31 July 1945.

You are obliged to keep some time available for the completion of the task assigned to Heinkel Ltd. by the military government. For this you will be paid RM 300 per month until the task is finished.

With highest regards,
/s/ Heinkel
ERNST HEINKEL COMPANY
Hirth Engine Works

Along with the termination letter from Heinkel, Von Ohain received an enormous final royalty payment in accordance with his contract. Royalties made him, as he said, "a rich man who couldn't buy anything." The payment was in Reichmarks, then worthless because of Germany's defeat. Von Ohain had more than 50 German patents by the end of World War II. He was stripped of the patent rights by the war. Von Ohain did not see Ernst Heinkel after 1946.

In midsummer 1945, Von Ohain was contacted by the U.S. Navy. They had taken over the Heinkel factory in Stuttgart and found the burned-out remains of the He S 011 engines. They were impressed with the highly advanced level of technology and development of the engines. In August 1945, Von Ohain and Max Bentele signed personal services contracts with the Navy to be consultants for "independent development of gas turbines" as part of the U.S. Naval Technical Mission in Europe. The Zuffenhausen plant was reopened under Navy control. It was a very welcome opportunity. As von Ohain had hoped when the engines were burned in response to Hitler's order, he could continue the design and development of the 011 engines. Under the surveillance of Navy technicians, 12 engines were built for performance evaluation at sea level and at the Munich altitude facility and for their eventual shipment to the United States and United Kingdom. Von

Ohain and Bentele wrote detailed reports on the engine and its components.

Von Ohain said, "The Navy was highly interested in the improvements of the 011 engine [that we made during the time of the contract] but delays prevented us from developing it further. We used hollow blades for the turbine successfully, and had made a number of improvements in the axial flow compressor, but before such improvements could be used it was 1946 or 1947." By this time, there were more powerful turbojet engines being built in the United States and Britain.

The Navy protested when the Air Corps officer returned and made von Ohain a preliminary offer to come to the United States. The Navy would not release him.

> 14 Sept 1945
>
> To: Whom it may concern
>
> Dr. von Ohain is needed here in Germany in connection with a Navy Project—namely, the production of 011 Jet Engines by the Ernst Heinkel AG of Zuffenhausen.
>
> His removal at this time would be of great inconvenience to us. Besides, the procedure appears to be highly irregular, as the transfer of such a person has to be arranged through the local and regional military government.
>
> Please see me at Heinkel plant before removing Dr. von Ohain.
>
> Clarke G. Minter
> Civ. Tech. US Navy

Von Ohain remembers that there was a Navy lieutenant who discouraged his move to the United States. He said, "First you have to work for us [the Navy]." Furthermore, he told him, "You wouldn't fit in America. You aren't 'Arctic.' I will see to it that they don't take you to cold America. You are here in a beautiful area. You feel much better here. In America you would have to be much tougher."

Von Ohain was puzzled. He had always wanted to go to America. "I was a little bit disappointed because in my youthful days I had read so many books [about America]. People [in those books] were all helpful and friendly. I thought America was the country of freedom and unlimited opportunities."

Some of von Ohain's 011 engines were shipped to Trenton, New Jersey, in 1947 for bench tests. When they were examined by leading American propulsion engineers, they said, "The He S 011 was far ahead of its time!" Today, two 011 engines are housed at the National Air and Space Museum in Washington, DC.

In February 1947, work on the engines was completed. The Paperclip quota was already filled, but the Air Corps wanted von Ohain and made him a formal offer to emigrate to the United States. Max Bentele was invited by the British Ministry of Supply to work with a German–Austrian team on the design and development of a tank gas turbine.

Von Ohain readily accepted the offer. He considered that this move was a long way from the teaching professorship he had always wanted, but he liked that he would be able to continue his work on the turbojet. There was little left to hold him in Germany. Cities were in ruins, people were homeless, hunger was everywhere, and the chance of finding a job researching aircraft propulsion seemed remote. Germans feared the specter of Russia, in its postwar power play, rearming and overrunning Europe in the near future.

Hans von Ohain's life was meticulously reviewed and investigated by authorities of the United States for evidence of any Nazi activity. He had never been a member of the Nazi Party. His spotless record was, in large part, because he had been kept isolated and intensely busy by Heinkel. Recruiting him for work in the United States was beyond question.

Following are the terms of the contract von Ohain signed. The original contract could not be photocopied here because of its condition. (The original was in both German and English.)

SPECIAL CONTRACT FOR THE EMPLOYMENT OF GERMAN NATIONALS WITH THE WAR DEPARTMENT OF THE UNITED STATES

Terms of employment made the 4th day of Jan 1947 by and between the War Department of the United States of America hereinafter referred to as the employer, and Dr. Ing. Hans von Ohain of Korntal hereinafter referred to as the employee.

1. Name von OHAIN (Last) Hans (First)
   a. Sex Male
   b. Place of original engagement (home address) Korntal Kreis: Leonberg, Saalplatz No. 2
   c. Position (job title) Arodynamics Engeneer [sic]
   d. Gross Pay twentyeight thousand Marks 28,000.00 per year

e. Minimum period of duty ____six month (6) year(s)
f. Name/address of person to be notified in case of emergency
Wolf von Ohain, Korntal, Gasthaus
g. Dependents (List name, age, relationship, and address)
Wolf von Ohain, 65, father, Korntal, Gasthaus
Katharina von Ohain, 58, mother, Korntal

The terms of the contract offered to von Ohain and the other Paperclip scientists required that they leave behind all personal possessions except 175 lbs of baggage. Scientific books and/or equipment evacuated with the employee were not charged against the baggage allowance. Travel was paid by the government to the location assigned, and employment was for a minimum period of six months, after which the employee could elect to return to Germany at the government's expense, or stay in the United States.

The employee's mission was to "undertake and prosecute to the best of his ability such scientific and technological tasks as employer may assign." Supervisors would be Army or Navy officers or civilian personnel. Work hours were the same as those for civilian employees performing similar work for the employer, with overtime compensated by time off. Necessary materials to perform the work, as well as scientifically and technically qualified assistants, would be made available.

The government furnished housing at no cost, "similar to that provided to junior officers and civilians," with incidental expenses borne by the employee. Government mess facilities were available for meals, with the employee paying the same as military and civilian employees. Medical care was free.

The employee would not receive the stated salary payment in the United States. It would be sent to a designated bank, or dependent, in Germany. The employee received $6 per day while in the United States and agreed not to send any part of such per diem to Europe.

There was a statement that there were no restrictions placed on religious beliefs, and "reasonable access" to recreational facilities would be available at employee cost. In event of the employee's death, local burial would be provided unless dependents requested that the body be returned to Germany.

The U.S. government had the right to determine who were dependents of the employee, usually wife and children. These dependents, upon the request of the employee, would be assembled at a location decided by the government. An effort would be made for the location to be in the German zone occupied by the United States, with relocation in the event of changes in the zone of occupation. A diet of 2300 calories per person with reasonable variety, and minimum fuel, would be available to dependents in Germany or would be augmented by the employer and paid for by the employee.

The employee could be returned to Germany to the original place of employment if an emergency arose with his dependents, if his health warranted, if his work was determined to be unsatisfactory from a professional standpoint, or if his conduct did not meet required standards.

Von Ohain signed the employment contract on 4 February 1947, and it was cosigned by a representative of the U.S. War Department. Henning von Gierke, another Paperclip scientist, shook his head when he recalled signing it and said, "We did not read the terms very closely." The 34-year-old Von Ohain was a single man. He claimed his parents, Wolf Sr. and Kitty, as dependents. Later he included as family his brother, Wolf Jr., Wolf's wife Hildegard and their baby.

Until passage to the United States was approved, dependents of the scientists were housed at Landshut, northeast of Munich, a complex where housing and food was provided by the U.S. government along with protection against possible recriminations from Germans who were not selected for departure. Von Ohain's parents, Wolf Jr., and Hildegard moved to a second-floor apartment in Landshut. Their downstairs neighbors were Wernher von Braun's parents. Altogether, there were 320 dependents living at the housing complex at Landshut.

Review of the scientists' education and previous positions in Germany is extremely impressive and indicates that there might have been a significant drain of knowledge by taking these individuals from Germany. Interviews with some of the scientists in later years are interesting in that they show great personal humility and national pride for the country they left. They admit that they were specialists in their disciplines, but they say that Germany had many more talented people.

The Germans gave their best efforts to contribute their work according to the contract they signed with the United States. A list of the estimated value of their work (in 1950 dollar values) is listed in Clarence G. Lasby's book, *Project Paperclip:*

Air Force development programs	$30,000,000
Supersonic wind tunnel research	Saving of five years
Rocket engine research	$6,000,000
Long range reciprocating engine research	$1,200,000

These figures do not include the yearly savings provided by individual specialists. The ultimate savings is estimated to be "nearly $2 billion."

There were some familiar names on the list of scientists who were at Wright Field in 1948:

–Hans Mauch, Hans Oestreicher, and Henning von Gierke were assigned to the Aero Medical Laboratory.

–Helmut Schelp and Gerhard Aichinger were assigned to the Equipment Research Laboratory.

–Hans von Ohain, Anselm Franz, Karl Volk, and Emil Walk were assigned to the Engineering Plans office.

–Otto Walchner and Heinrich Weigand were assigned to the Physics Research Laboratory.

Some scientists were sent on temporary duty assignments from Wright Field to another government location or to a civilian business where their particular field of expertise would be useful.

After the war, Hans von Ohain's parents remained in their home near Berlin until Hans discussed his opportunity to emigrate to the United States under provisions of Project Paperclip. He offered them the possibility to be his dependents and follow him to the United States when he had fulfilled his preliminary six-month obligation, after he had arranged housing. They agreed, left their home in Dahlem, and moved to Landshut until passage could be arranged.

A few years later their house at Dahlem was purchased by the Berlin Philharmonic Orchestra for use as a guest house for visiting artists. Later it was purchased by a doctor. Today it has been remodeled into a two-family residence by architects who live there. The house is hidden behind a high fence and tall shrubbery.

Herbert von Ohain and his wife elected to stay in Germany. Hans's brother Wolf worked for a short time as translator for the 7th U.S. Army Headquarters Document Center preparing for the Nuremberg Trial. He also accepted his brother's offer for him and his new wife to become Hans's dependents under Project Paperclip. The von Ohain family looked forward to their new lives in the United States.

Von Ohain arrived in the United States at Wright Field in February 1947. He was never interviewed about his travel from Germany to Wright Field in Dayton, Ohio. Descriptions from government archives can trace the journey.

Scientists came alone. They left their families behind at Landshut, waiting for arrangements to be made at a later time for them to join the men in the United States. Customarily a small group of scientists traveled together by train from Landshut near Munich, where paperwork was initiated, to Bremerhaven. The port was a processing center for military who had finished their tours in Germany and were returning to America. "Liberty" ships that carried the troops back across the ocean to New York took a few German scientists. Several hundred troops and the scientists shared accommodations in a huge windowless room in the belly of the ship outfitted wall to wall with bunk beds. Cabins on the ships were provided for wives and children of the scientists who had come to the United States earlier and had sent for their families. The small ships required two weeks of rolling and swaying, day and night, to reach New York. The scientists thought the food was good and there was always enough.

Emil Walk's son, Dieter, who was nine years old when he came with his parents to the United States, wrote this account of their arrival in New York:

> The ship came into New York harbor during the night and was anchored out a ways until morning for harbor clearance and a tugboat to assist us in coming in for docking. In the morning when I stood on deck, we saw a sight that every immigrant dreams about, hears about, and experiences with a feeling of awe and wonder.
>
> Only if you are a foreigner can you experience that arrival to the shores of America and the sight most often mentioned—the Statue of Liberty. This was an emotion-charged moment, being on the deck of the ship which was anchored a few hundred yards from this gigantic statue. It seemed like a long time that we stood there looking at that Lady holding the torch.
>
> A new beginning. Hope and a level of expectation that we only experience a very few times in our lives.

At the New York harbor the scientists, with their suitcases of meager belongings, were herded onto a bus that drove them through the famous city's caverns of tall buildings to Grand Central Terminal. They faced an overnight train trip of about 12 hours to Dayton, Ohio. From the Dayton railroad station, a blue Air Force bus drove them through the city's downtown and along Third Street to an in-processing center at Building 11 at Wright Field, and then to the complex of wooden barracks of the Bachelor Officers' Quarters (BOQ), where they would be housed. One scientist interviewed years later said that when he first saw Dayton, Ohio, as they came out of the railroad station and drove through the dismal neighborhoods near the air base in the early morning light, he vowed he would never stay. There were gray houses and small factories. He did not see green spaces and parks. Dayton had no mountains and clear blue lakes, no grandeur of German cities or the charm of German villages he had left behind.

At the air base's administration building, the scientists' baggage was searched. After they were fingerprinted and photographed, their identification passes were marked on the back with a large "S." They were assured that they were not being confined, but their comings and goings were monitored. A "custodian" was assigned who was responsible for "surveillance, administration, maintenance of personnel security dossiers, assignment of quarters, and security of the German-specialist personnel." All "specialists" were required to sign a roster twice a week in the Foreign Administration Office in Building 11.

The men were housed in the Hilltop barracks at Wright Field. Living conditions were not pleasant. The primitive wooden barracks were divided into "rooms" by sheets of plywood that did not reach the ceiling but allowed heat from the single coal stove to circulate in the wintertime. The uninsulated building was sweltering hot in the summer. A communal bathroom served them all. Several of the scientists complained bitterly about the living arrangements. It was embarrassing to them when a new group of scientists arrived and saw how their compatriots were living.

The scientists had to have permission to receive mail, and that correspondence could come only from persons authorized by the base commander. Outgoing mail, whether domestic or home to Germany, had to be sent "open" through the Foreign Administration Office where it was read and censored with scissors before being sent on. Criticisms of living conditions or assigned work were not acceptable subjects to discuss. Correspondents replying from Europe had to address letters to the scientists "In care of the Commanding General, Air Materiel Command, Wright Field, Ohio."

The $6 per diem pay was enough to buy meals in the mess halls on base, but each scientist carefully saved as much as possible in order to send CARE (Cooperative for American Remittances to Europe) food packages to families and friends in Germany. They were allowed to mail one package per week through any U.S. Post Office. Items were restricted to nonperishable foods, clothing, soap, and medicines. Von Ohain said dinner was frequently only a can of Campbell's soup warmed under the hot water in the bathroom of the barracks to save money because he knew of the hardships at home. When restrictions were lifted, he sent packages of sugar, coffee, and canned sausages to his former professors and even some former girl-friends.

United States citizens emerged from World War II affluent from wartime employment, but consumer goods were scarce. Refrigerators and washing machines for the parents, and mechanical toys such as tricycles and electric trains for the children, were only beginning to be built in factories where in the previous four years wartime materiel had required three eight-hour shifts of production. Returning servicemen who had endured battlefield stress and independence and had married and fathered children returned to the States to an acute housing shortage and moved into their parents' homes. Women who had held responsible wartime jobs were laid off to give work opportunities to veterans and were expected to be happy staying home and having babies. The comforts of home and family were emphasized.

Communist Russia was under suspicion by the Western world. Winston Churchill made his Iron Curtain speech, saying "An iron curtain has descended." President Truman ordered a loyalty investigation of all federal employees, and the House Unamerican Activities Committee began investigations. The United States adopted a policy of "cold war" to contain Russia and Communist influences, and the Berlin Airlift of necessary food and goods was initiated. The government allocated large sums for the rebuilding of Europe through the Marshall Plan and other programs. Men whose education had been interrupted by serving in the military returned to college under the provisions of the G.I. Bill, which paid tuition and some living expenses. Very popular subjects were engineering and science, studies of the new technologies developed during the war.

Chapter 13

Research and the U.S. government

Research was to be Hans von Ohain's professional life in the United States in the next 32 years with the Air Force and for several more years after his formal retirement. Research is an art practiced by a person skilled in a discipline. It is a creative technique applied to a problem. Hans von Ohain had creativity to a superlative degree. With the Air Force, he was the right person in the right place to improve propulsion through his own work and through his motivation of others with similar talents. His achievements over the last 50 years of his life add up to enormous improvements in propulsion technology, one small piece at a time. Von Ohain, inventor of the turbojet engine, had the opportunity to see his invention develop to high complexity and widespread use.

Research, in the basic sense, includes the creation of new knowledge, the reorganization of old knowledge, and the establishment of values. "Fundamental, basic research" is largely theoretical, pursued by physicists, mathematicians, and other specialists. Its result is most often a written report, but a prototype for illustration may be produced. It is made up of many small improvements that do not result in any single soaring achievement. A major invention is rare. Research requires a steady creative process, looking at every element of a design to find a possible flaw or a recommended change.

Research and development (R&D) uses the new information, but may carry it further into a practical item. This effort requires both creative specialists and engineers. "Development," used alone, generally is an improvement to an existing item, and engineers take responsibility with hands-on work. "Applied research" is a synonym for R&D. There are no clear-cut boundaries for these categories. A laboratory such as those at Wright–Patterson may require input from several research disciplines for a single project. Universities traditionally are expected to be centers for performance of basic research. Small specialized businesses or small departments within major businesses provide R&D solutions under contracts with government laboratories.

It is difficult to justify funding or to get public support for basic research, that is, the discovery of information not yet known. Solving problems for a particular situation is considered to be more acceptable. The press is quick to report that the government is financially irresponsible when an expensive research project seems frivolous. Research carries the risk that many projects initiated may never produce a useful result. A finding or improvement may lie dormant for years. Financial managers without an understanding of the value of research become upset if a useful result comes from only one in ten projects, yet examination of expenditures will show that basic research consumes only a small part of the government's research and development, test and evaluation (RDTE) dollars.

For national survival, governments need research to explore new possibilities, new principles, and novel weapons, and become aware of technologies developed in other countries. The United States recognized the important role of air power during World War II. Appropriations increased for the National Advisory Committee for Aeronautics (NACA) during the war years from about $4 million in 1940 to more than $38 million in 1944.

At the end of World War II the Army Air Corps could not continue modifying and reconfiguring the planes from the 1930s that it had used for World War II. They took unique opportunities to learn the technical knowledge of their former adversaries in Germany and to examine new technology in the United States that had been postponed by the war. Problems for establishing research laboratories following the war included funding, staffing, and adequate facilities.

At first Congress provided only about half the funds requested for the government's proposed plans for research, perhaps reflecting the apathy of their constituents toward things military due to the war's end. Congress had not yet connected with the plan to bring German scientists to the United States and may have believed that the Air Force could not find the scientific talent to staff the projects. The aircraft industry could not be depended on for research because they were short on funds, paralyzed without the profits from procurement dollars that had funded R&D during the war, and purchases had dried up with the return to peace. General Electric almost stopped their development of the jet engine.

Veterans of the war were assigned to supervisory positions in the laboratories. They were well trained in tactics, but they were not educated in technical specialties. Many lacked skills in objective analysis and critical thinking or in reaching conclusions only when warranted by adequate evidence. They had little understanding of the projects for which they made decisions and were unable to evaluate contributions from universities. Military academies changed curricula a few years later to remedy this inadequacy.

The U.S. government was finding it difficult to hire qualified scientists because of the inadequate laboratory facilities at Wright Field and the competition of private industry's generous salaries. They proposed finding men with "the true research scientific mind, less concerned with personal recompense." The Paperclip scientists fulfilled that criterion, but their supervisors were not sure what to do with them when they arrived in the laboratories. They were better educated and more advanced in their work than most of those already working there. They were held to minor advisory roles in spite of the high positions they had held with major companies in Germany. Their only responsibility was to prove themselves over a period of time. Their one satisfaction was that their pay was based on their educational level, generally advanced. Some civil service people made the Germans feel unwelcome. They thought that the Germans would steal Amercian technical secrets, take them to Germany, and use them against us. They were polite as they worked together, but made negative remarks behind the scientists' backs and rejected them socially. Propaganda of hate had been effective on both sides.

The laboratory's process of outlining a project and its specifications in a statement of work (SOW), looking for a contractor to perform the work, and waiting for a report to be submitted was puzzling to the arriving Paperclip scientists. They were accustomed to performing the work themselves, and the facilities and laboratories were not as well equipped as those in Germany had been. Authorization for funding for basic supplies was slow. Equipment shipped from Germany arrived in poor condition with no immediate funds to repair it. The German scientists decided that their potential for contributions to Air Force efforts was wasted while they sat around doing nothing. Some applied immediately to return to Germany, but most stayed in the United States. In spite of initial discouragement, they believed that life would be better for their families in the United States. In Germany, recovery was slow and there was no work to compare with the opportunities and professional freedom at Wright Field. The scientists, recruited from all over Germany and Austria, enjoyed getting to know each other in the United States. Friendships and professional associations lasted throughout their years in Dayton and as the scientists moved to other locations.

Wright Field in 1946 should have been the ideal location to send the well-educated and experienced German scientists with specialties in the aeronautical sciences. Air research for the United States had been centered at Wright Field since the Army established the area as McCook Field in 1917, with its first research done in the Airplane Engineering Department. Extensive R&D in all aspects of flight continued during the 1920s and 1930s. Facilities were placed for studies in all disciplines, although during World War II improvements and "fixes" had been the priority over research.

In November 1943, the Jet Thrust Propulsion Laboratory was established at Wright Field after the Whittle experimental engine was acquired from Britain. Engineers provided input into the development of the Lockheed P-80A Shooting Star that became operational in 1945. Jet engine technology became an urgent learning process for all disciplines. In the 1989 Aeronautical Systems Division History Office publication, *Wright from the Start,* Jim Aldridge wrote that

> the arrival of the jet engine transformed at a stroke a technical community grown comfortable with a mature propulsion technology and the air vehicle built around it. With the new means of propulsion, not only the power plant but every aspect of the airplane was affected: its aerodynamics and hence airframe, including materials and structures; landing gear; radomes and canopies; fuels and lubricants; electronics and controls; cockpit and ejection systems. Jet aircraft raised a host of new problems for aero medical research. They opened up whole new areas of operations as well, as aircraft were now capable of flying higher and faster than ever.

Wright Field was home to the Air Document Division for material taken from Germany after the war. The material found cached in dry cisterns and salt mines throughout the defeated country was inventoried at the Air Document Research Office in London, and nontechnical documents were separated out. The documentation and related equipment were then sent for analysis and exploitation to the Air Document Division established at Wright Field and grouped by subject: aerodynamics, engines, structures, electronics, rockets, and miscellaneous. The material had a wealth of information that could be provided at once to Wright Field laboratories.

On 13 January 1948, Wright Field joined its next-door neighbor, Patterson Field, to become Wright–Patterson Air Force Base.

Air Materiel Command at Wright–Patterson organized a research unit in February 1948 named the Applied Research Section, funded as part of the Engineering Division, Directorate of R&D. They received some funding from the Congressional budget, expanded some laboratories, and selected a few projects. The chief scientist was Arnold Kuethe, well known for his textbook, *Foundations of Aerodynamics,* used for many college engineering classes. Supervisors were Elmer Johnson and Roscoe Mills, both with years of experience building wind tunnels during the war. Hans von Ohain's career with the Air Force began as a bench-level scientist in the Analysis Division of that laboratory.

Even before he left Germany in 1947, von Ohain had been given a position description for the work he would do for the United States Air Force: "Mechanical Engineer. Conducts theoretical work on advanced airbreathing propulsion systems and test facilities," it said.

The position description was unchanged through 1952. It combined his great love, physics, with his second love, turbojet power. He described his job as dealing "specifically with problems of high temperature air generators and compressors." He was not a civil service employee, but worked under a personal services contract, as did the other Paperclip scientists.

The Applied Research Laboratory was subsequently renamed the Flight Research Laboratory (FRL). The laboratory pursued research in the fields of aerodynamics, applied mathematics, applied mechanics, chemistry, metallurgy, physics (including nuclear physics), propulsion, special projects, and computation. The director of the Air Force Office of Scientific Research in Washington, DC, believed that the mission of FRL should be to select topics to put under contract with academic specialists at universities. FRL personnel would be responsible for only "quick fixes," not performance of in-house research efforts. The highly qualified specialists quickly protested that in order to be well advised they needed to keep abreast of all new developments, requiring their own hands-on work. The management at FRL quietly made in-house research possible, even if the efforts were "underground" until it was safe for them to be publicized. The policy of sending the scientists to national and some international seminars was established. The skilled scientists in the laboratory, including the Germans, provided unique contributions.

Von Ohain was left mostly on his own to continue with work he had begun in Germany. He did not let any lack of equipment or administrative assistance frustrate him. He was accustomed to working within his own mind, and he had many ideas he had not had time to write down during World War II's urgency. His Air Force employers were dazzled by his knowledge and by the contributions he easily made to the technology they were exploring. Stumbling to express himself with little knowledge of the English language, von Ohain was pleasant and cooperative and earnest with his American management.

Von Ohain could not recall the name of his first supervisor, but he described him: "He had an Irish accent. He was fantastic. He would go through the office in a very friendly way. He said, 'Can you give us ideas that you had?'" Exploitation of the Paperclip scientists began by having them write down everything they had worked on in Germany. Hans said:

> I wrote a number of reports. I said we had a new engine principle which I described with great pleasure because it was what I had in mind, and we did a little for the Navy already. The Navy sent a small engine based on the static pressure exchange principle. I was angry when one guy gave a negative view at a meeting about Project Squid for the Navy. We had one like that in Germany but could not get strategic materials.

Two important national leaders had influence on von Ohain's professional life in the United States, the commander of the Army Air Corps, Gen. Henry "Hap" Arnold, and the famed aerodynamicist, Theodor von Kármán. The vision of these men blessed the future of United States air superiority late in World War II and the years immediately following.

Gen. Arnold was trained to fly by the Wright brothers, becoming one of the first two active pilots in the U.S. Army. He had a series of "firsts" in his early flying career. His vast flying experience earned him a desk in Washington, DC, during World War I. In August 1917, at age 31, he became the Army's youngest colonel. In the years between World War I and World War II, he became an advocate for the Army Air Service, later named the Army Air Corps, and held a series of increasingly important positions. During World War II, Arnold became commanding general of the Army Air Forces, liaison with the British Chiefs of Staff, and a full member of the U.S. Joint Chiefs of Staff. As a result of Arnold's lifetime of leadership, the Air Force became an independent branch of the U.S. armed forces in 1947.

In 1941, Gen. Arnold learned of Frank Whittle's radically new turbojet engine and made arrangements for it to be brought to the United States. In November 1944, even before the end of World War II, Arnold deduced that the war was being won by the Allies more by sheer force and mass production rather than by technical superiority. He believed that this scientific deficiency should never be allowed to happen again. He assigned Theodor von Kármán to study existing scientific achievements of both friend and foe and to forecast, if possible, those areas that would be most promising of future applications. "What I am interested in," Arnold said, "is what will be the shape of the air war, of air power, in five years, or ten, or sixty-five. Gather a group of practical scientists for all the new things. I want to know the impact of jet propulsion, of atomic energy, of electronics. Make me a report." He wanted to anticipate and lead future requirements rather than waiting for them to materialize.

Von Kármán was born in Hungary in 1881 and educated at the University of Budapest and Göttingen University in Germany. He won world acclaim for his research in mathematical analysis, thermodynamics, and aerodynamics, and he was a pioneer in high-speed aircraft and missiles. After an outstanding career as teacher at Budapest, Aachen, Göttingen, and other locations, he came to the Guggenheim Aeronautical Laboratory at the California Institute of Technology (CalTech) in the United States. He became an American citizen in 1936. He acted as advisor to many international scientific boards and industries and received honors from many scientific organizations across the world.

Von Kármán assembled a group of specialists from the advisory group of which he was chairman and prepared the special studies. Information was gathered from both Allied and enemy research centers and from scientists including Ludwig Prandtl of Göttingen, Walter Dornberger and Wernher von Braun of Peenemunde, and Hans von Ohain of Heinkel. It resulted in 12 volumes named *Toward New Horizons* that outlined ideas and goals for the future development efforts of the United States. One study, *Science, the Key to Air Supremacy,* authored by von Kármán emphasized the importance of funding for research to enhance the technological base of the United States. It included a review of the work of German scientists during the war and encouraged their recruitment. Von Kármán concluded that the technological advances made in Germany during World War II were "not because of superior scientific personnel, but were the result of substantial government support." He recommended that future R&D funding account for one-fourth to one-third of the peacetime Air Force annual budget. This recommendation was made, according to one comment, "optimistically, if not naively." In late 1945, the studies prepared by von Kármán's committee were reviewed by the Scientific Advisory Group and reported to Gen. Arnold. The report used "funny words" that are now accepted as though they had always been a part of our language: turbofan, turbojet, turboprop, pulsejet, etc. The report was classified secret and was not declassified until 1960.

The recommendations in the reports provided impetus for men who had strong opinions about the value of basic research and also were familiar with the facilities at Wright Field: Gen. Donald L. Putt, Gen. Franklin Carroll, and Frank Wattendorf. They believed that Wright Field should be built up to be the center of Air Force research.

Gen. Donald L. Putt was a powerful leader and a strong promoter of fundamental research. He had been a project officer in 1945 in the Air Technical Service Command (ATSC), the organization responsible for collecting, analyzing, and preserving German technical information. He helped find and select scientists for the Paperclip recruitment and evaluated the value they might provide. He was acquainted with the name and reputation of Hans von Ohain from reports during his time in Germany. While he was a colonel at Wright Field he learned the capabilities available there. He was promoted to general in 1949, based in Washington, DC, as director of R&D in the Office of Deputy Chief of Staff and as military director for the Scientific Advisory Board. His influence for basic research continued in 1952, when he became commander of Wright Air Development Center. As a lieutenant general in 1953, he took over Air Research and Development Command (ARDC), an organization above the Office of Scientific Research. In 1954, he returned to a staff position at the Pentagon.

Based on his own experience and the von Kármán study, Gen. Putt believed that in peacetime, R&D should have top billing in the Air Force mission. In 1949 he wrote that there should be assurance, even requirements, that new developments are produced, operationally tested, and used. R&D funding should not be influenced by procurement, that is, the actual purchase of items or ideas for use. Long range values should be a major factor. He declared that all research activities should be unified under a newly established deputy chief of staff for R&D, and research policies should be disseminated from a high level. The publication of his statement led to action at a high level. An Air Force policy was published affirming that men, money, materials, and facilities were to concentrate on "new and radically improved materiel and techniques" and to limit work on existing types of aircraft and armaments.

Gen. Franklin Carroll was with ATSC at Wright Field in 1945 as chief of the Engineering Division. He returned in 1949 to be Air Materiel Command's director for R&D. He recognized the long heritage of technical innovation at Wright–Patterson Air Force Base and advocated expansion of its facilities.

Frank Wattendorf had degrees from Harvard and Massachusetts Institute of Technology (MIT) and studied further at Göttingen University in Germany with von Kármán and Ludwig Prandtl. He returned to the United States as an assistant to von Kármán at CalTech. He designed and supervised the construction of China's first large wind tunnel and the 40,000-horsepower wind tunnel and other research and test facilities for the Army Air Force at Wright–Patterson Air Force Base. He traveled to Germany in April 1945 and discovered the Germans' huge sonic wind tunnel at Oetztal under construction. At Kochel he found supersonic wind tunnels in operation and a hypersonic wind tunnel under construction. Wattendorf was employed as research advisor to the Engineering Division in the laboratories at Wright–Patterson when he became a member of Gen. Arnold's Scientific Advisory Group (later, Scientific Advisory Board). Within a few years Wattendorf provided outstanding guidance in developing the Arnold Engineering Development Center. Later he helped establish the Advisory Group for Aeronautical Research and Development (AGARD) and became its first director. AGARD, part of NATO, was chaired by von Kármán.

While von Ohain was considered at FRL to be a good worker and a very intelligent man, there were other German scientists who were described the same way. Suddenly von Ohain was "discovered" at Wright Field. Cliff Simpson, a supervisor in the Power Plant Laboratory, had interviewed von Ohain in Kolbermohr before the end of the war. He told Frank Wattendorf about the value of what von Ohain had done in Germany, and Wattendorf passed the information to von Kármán and Gen. Arnold on the Scientific Advisory Board. Von Kármán recognized the superiority of von Ohain's Göttingen education and knew of the creation of his turbojet engine. Von Kármán and Wattendorf wasted no time in bringing von Ohain into their planning as a member of the Aero Facilities Group.

In March 1948, Gen. Carroll and Wattendorf formed the Aero Facilities Group at Wright–Patterson to plan an advanced testing facility for the Air Force. Von Kármán led the planning from an office in Washington, DC. Roscoe Mills and Elmer Johnson became team leaders. Von Ohain was recruited to study the future compressor requirements. Two brilliant young scientists, Maurice O. Lawson and Anthony Fiore, also were recruited. Close working relations and friendships developed among these men that lasted many years. An impressive list of scientists was assembled from international circles on this side of the Iron Curtain as advisors to the planning committee. Formal planning of the facilities began.

Von Kármán was a frequent visitor to the planning committee at Wright–Patterson. He disliked staying in hotels. When he came for meetings, Mills would rent a hospital bed and put it in his office. There was a restroom with showers in the building. Von Kármán stayed in Mills's office for as long as a week using the office restroom and eating at the base cafeteria.

The most urgent need for the Air Force was a state-of-the-art wind tunnel to exploit sonic- and supersonic flight regimes rather than farming out the testing to commercial companies. From the early 1940s, Wright Field was Gen. Arnold's first choice as the location for the testing center and some new facilities, including a 10-ft sonic high-altitude wind tunnel, a 2-ft supersonic tunnel, and a 12-ft spin tunnel. Some advanced facilities for research had been constructed in the early days of flight development, including a 14-in and a 60-in wind tunnel. In 1942, under the leadership of von Kármán, Wattendorf, and MIT's John Markham, a 20-ft, 40,000-horsepower wind tunnel had been built, the largest in the world at the time, that could test aircraft up to 450 mph. Wattendorf led in the building of another wind tunnel with a 10-ft section capable of transonic speeds. In 1944, a supersonic tunnel was built. Arnold's plan for additional facilities had to be laid aside until the war ended.

Wattendorf, in his position with Gen. Arnold's Scientific Advisory Group, had realized in 1945 that Wright–Patterson and the Dayton, Ohio, area did not have the electrical power, the water supply, and the required space for the test facilities identified by their plans. An enormous quantity of stored hydraulic power was necessary to run the multithousand-horsepower motors for the big wind tunnels. Wattendorf formalized his opinion in a document, the "Trans-Atlantic Memo," written when he was traveling home from gathering information in Europe. He recommended finding an area where major power would be available.

The decision for the location of the new facility required years of political maneuvering among members of Congress. Grand Coulee Dam on the Columbia River, 90 miles west of Spokane, Washington, was the largest dam in the United States, but there was opinion that its location was vulnerable to foreign attack from the coast. Hoover Dam, formerly Boulder Dam, on the Colorado River in Nevada was another strong contender. Those states' congressmen lobbied long and hard to have such a major facility in their own districts. As a politically acceptable alternative, southern middle Tennessee at Tullahoma was selected because of its availability of land, water, and power from the Tennessee Valley Authority.

The process for Congressional approval of funding began. Von Ohain helped with the planning and waited with his peers for the decision to proceed with the actual building of the testing center.

In March 1950, after long delays associated with the legislative process, the planning team of Wattendorf, Carroll, Mills, and Johnson got approval to build the initial nucleus of the Air Engineering Development Division. Gen. Frank Carroll was named commander at Camp Forrest, answering to von Kármán, and Frank Wattendorf became his deputy chief and scientific advisor. The facility would have a 16-ft-square transonic-supersonic propulsion wind tunnel with capability up to Mach 3.5 and systems brought from Germany: the Kochel battery of wind tunnels capable to Mach 10 (later dedicated as the Von Kármán Facility) and the fully operational BMW jet engine test facility (known as the Engine Test Facility or ETF).

Von Ohain and the other members of the team had to make many temporary duty trips to Tullahoma to assist in the building of the facility. Von Ohain often traveled with young Lawson, known as "Mo," who fussed at him to practice his English vocabulary. They traded English and German language lessons during work and social times, both serious and casual, and became close friends. Their wives remain very close friends today.

The completed facility was formally named Arnold Engineering Development Center (AEDC) in honor of Gen. Arnold's vision and aggressive policy leading to the advanced facility. It was dedicated by President Harry Truman on 25 June 1951. Truman promised, "Never again will the United States ride the coattails of other countries in the progress and development of the aeronautical art." Since its establishment, AEDC has tested virtually every high-performance aerospace system in the Defense Department's inventory. It is used by both government and the aerospace industry.

Von Ohain returned to FRL in 1950 to work on compressors in Building 190T under the supervision of Elmer Johnson. When the Communists swept down from North Korea and the United States rose to assist South Korea, funding was cut for FRL. President Truman made a statement about setting priorities, and the Air Force hurried to move funds from research to improving the existing fighters in production, reverting to its World War II policies.

With Russia's announcement of their testing of an atomic bomb, the United States feared an impending war. Sen. Joseph R. McCarthy began investigations of communist leanings of Hollywood actors, proceedings that were widely publicized on the popular new home appliance, television sets. The Korean conflict required 250,000 American men for military service, and the war was going badly. The conservative, comfortable populace looked for a candidate to replace Harry S. Truman for the next presidential term and turned to Gen. Dwight D. Eisenhower for future leadership for the country. His platform was appealing: "K1C2, against Korea, Communism, and Corruption."

A committee in the Air Force's Scientific Advisory Board persisted in their belief in basic research. They urged the establishment of a new Research and Development Command, with an office at Air Staff level to oversee its efforts. In the midst of the Korean War, ARDC was established. It was the first major air command wholly devoted to R&D. Facilities were expanded to conduct basic research on a laboratory level. Hans von Ohain's position in FRL was funded by Wright Air Development Center, part of ARDC. Along with adding funding for research to the budget, the government published a statement drafted by Gen. "Jimmy" Doolittle that tried to resolve a long-standing argument between engineers doing research and those doing development. It read:

> **Research,** which is devoted to long range goals, requires a constancy of purpose which can be secured only by protecting research from the urgency of immediate engineering problems arising in connection with the development function.
>
> **Development** is a function which runs continuously from the concept of an idea to the obsoleting and abandoning of the process or product that resulted from the idea. Frequently by far the greater part of development takes place after an airplane is in actual service.

Chapter 14

Family life

Hans von Ohain had remained unmarried after college and during his years with the Heinkel Company. While socially charming, he was intensely occupied by his work. The demands of his position at Heinkel and wartime urgencies required long hours at the factory. When he arrived at Wright Field, he found he was expected to work only eight hours a day, five days a week, and a half-day on Saturday. For the first time in his life, he had time for recreation and social life.

Within a few months after their arrival at Wright Field, the German scientists were authorized to travel on weekends within a radius of 50 miles of the base without special permission, and they were granted leaves of absence of up to 72 hours. The men appreciated the freedom to explore the new community. One scientist remarked that the YMCA membership provided wonderful outdoor recreation. For 25 cents one could sign up, he said, for "gasoline transportation" (a bus) to go with a group on Sunday afternoon to Glen Helen Nature Preserve or the little towns of Yellow Springs or Germantown to walk around. "You were called by your first name right away, and no one asked where you came from. We had friends right from the beginning."

Gradually, based on individual histories of personal conduct, trustworthiness, and compliance with regulations, the scientists were allowed to move out into the community to live in rented rooms and apartments. Because there had been no home construction during the war, they, along with the returning veterans, faced a housing shortage. The scientists wanted to bring their families from Germany. When they finally were able to find a place to live in Dayton and brought their dependents from Germany, their children helped them fit into the community, attending school and playing together with American children. Von Ohain, the bachelor, was able to find a room to rent in Dayton on Emerson Avenue in a good residential area.

In the 19th century, a large settlement of German immigrants had settled in central and southwestern Ohio. German accents could be heard in many neighborhoods and German surnames were very common. Dayton was surrounded by farms, and farmers brought their fresh fruits and vegetables to the downtown Market House on Main Street on weekends. While shopping there, the German scientists found one of their favorite foods that they had been missing: *wursts* (sausages, lunch meats). These were traditional German "lunch meats," including *leberwurst, blutwurst, teewurst, sulze, schinkenwurst,* all varieties, along with "German bologna" and frankfurters. They learned that the *wursts* were made by a man named Albert Schukat who had a small grocery at 301 Xenia Avenue. His store became very popular with the new German residents. In the same neighborhood was a traditional German bakery, Schattschneider's. The Germans began to feel at home.

The Schukat family lived on a 120-acre farm north of Dayton near Brookville. Albert Schukat had come from Germany with his family in 1929. The Schukats had no children of their own, but years before they had adopted a baby from a family whose mother had died in childbirth. The Schukats had given up a comfortable home, where they had servants, to come to America. Friends in Dayton invited them to stay with them until they could establish their own home. The Schukats found that the United States was in the depths of the Depression, but, luckily, they found employment as servants to a wealthy couple. They saved their money and bought a grocery store and saved again and bought a farm. In those first years in America it was lonely for their young daughter until they provided her with a companion, a German shepherd dog like one she had had in Germany.

Mrs. Schukat was a warm-hearted woman who understood the loneliness of the homesick Germans scientists. Every Saturday she baked a cake and on Sunday afternoons the German scientists they had come to know were invited to the Schukat farm for coffee and cake. It was not until months after he arrived in the United States that Hans was invited by his friend, Otto Walchner, to go with him to the Schukat home.

At Christmas dinner in 1947 Hans met the Schukats' daughter, Hanny, now grown up. She was a pretty, cheerful young woman and a talented singer who had studied at the Cincinnati Conservatory and in New York City. She had boldly gone to New York by plane, a DC 6, at a time when train transportation was conventional.

Hanny played light German songs on the parlor piano—Lehar, Kalman, and Strauss. She sang and loved to dance. Hans was charmed. He went back to the Schukats' home many times in the following months. Hanny had dates with some of the other German scientists, and Hans politely did not interfere with her social life. The Schukats enjoyed entertaining the Germans but particularly liked Hans, who was warm and friendly and entertaining with gentle humor.

Pretty Hanny Schukat had her eye on handsome young Hans von Ohain who, along with his German friends, often visited her home on Sunday afternoons for cake and coffee. She took the initiative and asked Hans to escort her to her church choir picnic in May 1948 at the home of her friend from Music Club, Edna Jacobson. Her father and mother were delighted when Hans and Hanny fell in love. Hanny was an eminently suitable choice as a wife for Hans von Ohain. She was attractive, had an engaging, sweet personality, and came from a cultured German family. Hans was contented, having a pleasant social life and the luxury of freedom to pursue his own thoughts at work. They became engaged a year later.

Hans von Ohain and Hanny Schukat were married in November 1949. Otto Walchner, a Paperclip scientist and friend of Hans from Göttingen's Aerodynamic Research Center, was best man. The Schukats had what Hanny called a "very nice" party at the farm. On the day of the wedding Hanny and her family were startled to find out from Otto Walchner that Hans had an international reputation as the inventor of the first turbojet engine that flew. In the almost two years they had known each other, Hans, with his usual modesty, had not seen a reason to tell them about his work on the jet engine or his place in history. Perhaps he wanted a partner in life not for his fame and possible fortune, but someone who loved and wanted him for everything else he was. Sigmund Harder, their friend for many years, described Hans and Hanny's relationship through life: "It was based on total trust, mutual respect, acceptance of each other, and total dependence upon each other, interwoven with their love."

Hans wrote home to his family in Landshut about his marriage. They were very surprised that he had married. "We thought he was an old bachelor," his brother said. Hans explained that they would be living at the Schukats' home on the second floor of the large farmhouse in order to save money to bring Hans's dependents to the United States from the apartment at Landshut. In a letter home, Hans tried to describe the large and well-kept farm. Because the family had seen the movie *Gone with the Wind,* they knew it would be like Tara, Scarlett O'Hara's home.

In the week after Hans and Hanny's honeymoon, and over two years after he had arrived in the United States with Project Paperclip, Hans's entry into the United States was made legal. It had become awkward for the scientists who had married since arriving, or had children born in the United States, to have no record of having entered the country. In November 1949, Air Force authorities drove a busload of German scientists, including von Ohain, from Dayton, Ohio, to Niagara Falls, New York. After crossing the border into Canada they received visas and walked back to the United States across the Rainbow Bridge as legal immigrants. Hans would have to wait almost five years after this date before applying for American citizenship.

Life at the Schukat–von Ohain home was warm and loving and busy. Hanny's parents lived downstairs in the farmhouse, and Hans and Hanny, upstairs. While the grocery store in Dayton was his principal interest, Albert Schukat recalled his experience of managing a dairy farm in Germany and resolved to be the best farmer in America. "He tried anything," Hanny said. The Schukat farm was typical of large working farms in the early part of the 20th century. Albert drove the tractor and prepared the farmland for spring planting with the necessary plowing and disking. He grew wheat, corn, oats, and hay for feed. The farm had livestock: cows, chickens, ducks, and pigs. After butchering was done at the next farm, Albert made smoked hams and filled casings, *wursts* for his grocery store. There was an extensive vegetable garden for the family's use, with enough to take to market. The land is no longer farmed. It has become part of Sycamore State Park at the corner of Diamond Mill Road and Wolf Creek Pike. The spring still yields clear, clean water. In later years, one of the Schukat barns was preserved by being dismantled and taken to Hueston Woods State Park as a historical exhibit.

Hans loved to live at the farm, but he never considered becoming a farmer because he was fully occupied with his work at Wright Field. Hanny reported that he did not drive any farm equipment or help with the harvests, but he helped "Opa" Schukat milk the cows every morning before he left for work. Hans enjoyed the outdoors and liked to take long walks through the farmland just as his parents with his brothers had done through the German forests and parks, especially on Sundays. He found that at the farm in Ohio he had to walk along the side of roads because fences divided the farms into small fields, while in Germany he had been able to walk long distances across fields without fences. In Germany, forests were cleaned of underbrush to allow tramping through the trees and only pastures were fenced.

Hanny and her mother were kept busy taking care of the house and the children: young Steve, Hanny's son from a previous marriage; baby Christopher, who was born a year after Hans and Hanny were married; and Cathy, born three years later. Hanny helped her parents with the work of the farm and the grocery. She loved music, and in the evenings after the children went to bed she sang and played the piano. Hans enjoyed hearing the classic German lieder and the light operetta songs. Hanny took voice lessons and sang solos at a neighborhood church.

Seven-year-old Steve enjoyed his new papa, another grown-up in the house, who treated his mother so kindly and took her to nice places on Saturday nights. Steve liked to listen to him talk about things and was sure he was the smartest man in the world. When he admitted this to his classmates at school one day, they all groaned. "Everyone thinks that about their dad," they said. They did not realize his papa was Hans von Ohain.

Hans could speak German every day. His wife and the Schukats spoke German. Hans and the other German scientists at Wright Field were determined to communicate with each other in English even though the words were slow and stilted. It was up to them to develop sufficient English to communicate with the American scientists. At the least they hoped to achieve what von Kármán called "the universal scientific language, broken English." Hans said he was grateful that he had studied Latin because it helped him with English vocabulary and grammar. He and his new wife attended many American movies and watched television because he needed to learn slang and how to understand American jokes.

Hans was determined to bring his parents to the United States as soon as he could find housing. The $5000 per year he was earning under his work contract had to be spent very carefully. The couple's goal was to put a down payment on a house for Hans's parents and his brother and wife, Wolf and Hildegard. In the fall of 1950, Hans and Hanny von Ohain found a small house off north Salem Avenue in Dayton that could be adequate for his family from Germany. It was tiny and far from luxurious, with only two bedrooms, one of which was over the garage and unheated. In November, Wolf and

Kitty von Ohain, along with Wolf Jr., Hildegard, and their baby, came to the United States. They were able to bring some furniture from the Dahlem house. They received a small pension from Germany that paid for their groceries. Hans's father appreciated being in a safe and warm home with plentiful food. Hans's mother, who had been born into the affluent Nagel family and had lived in beautiful Dahlem, had a difficult time adjusting to living in Dayton. Hans's brother and family lived with the elder von Ohains for a few years and had to use great patience. Hans was grateful that the elder von Ohains did not live with him and Hanny. When the von Ohain home in Dahlem finally sold, Hans's parents were able to purchase a better home in Lebanon, Ohio, and Wolf and Hildegard moved to the same town.

In July 1952, under a "permanent executive appointment," von Ohain and other Paperclip scientists in the laboratory became civil service employees for the U.S. government. This brought a pay increase and the potential for promotions. Von Ohain was no longer a contractor. In December of the same year von Ohain became an American citizen. He had spent time studying the required lessons and working on his English language skills. He said that he was proud to be a citizen of the country that produced Thomas Edison and the Wright brothers.

Happy years followed for Hans von Ohain with his family. It was almost a 45-minute drive between Wright–Patterson Air Force Base and the farm near Brookville. Hans left for work by 0700 hrs before the children were awake in the mornings. The school bus stopped for Steve, who went to Brookville High School, and Christopher, who had now started elementary school.

After school, Chris led young Cathy around the farm into all kinds of mischief experienced only by farm children, exploring the woods and extensive outbuildings. Big brother Steve and his friends sometimes played pranks. A tunnel ran under the hill, from the basement of the house to the road, an ideal place to lock little brother Chris and call it a dungeon. In the wintertime, Hans and the children walked across the fields to a big hill for sledding or skiing.

The children would meet their father excitedly when he arrived at home each night. They reported that their father's first kiss was always for Mom, followed by a kiss and hug for each of them. Cathy liked to slip her hand into her father's and lead him out to take a walk before dinner, a time she treasured as her very own time with him. Walks with their father were special to each child even when they were older and home from college.

Dinner around the big table was a happy time with everyone sharing their day's activities and their father listening to each one. Hans always had mischievous humor and teased Hanny in some gentle way. After dinner Hans watched the news on television. Cathy remembers how her father liked to tuck the children into bed when they were little. He would make an elaborate drape of a sheet over the head of the bed under which the children could hide. He called it their "angel bed." He sang them lullabies, and just before he turned off the lights he would slip little pieces of candy bar into their hands. Then he would disappear into his office in the house, or to the workshops he had set up in the basement and the back of the chicken house, to work on projects of his own, projects other than those at the base. The children enjoyed hearing him in his workshop, whistling and singing. When his "science buddies" from the base joined him, they heard machinery grinding and loud explosions. Hans liked to build his own machinery for experiments. As busy as he was, the children said their father was always approachable and could be interrupted for homework help.

Hans had ideas for a small compressor fan and a jet car and made a model of a pencil-sized toothbrush that automatically squeezed toothpaste. This never became a commercial venture as did the device he and Mo Lawson patented, the "Fill 'n' Flator," which could fill a child's plastic swimming pool with air and water at the same time. Hans always had a project for his evenings. He built the young children a tree house. On the wood floor of the big bank barn, Hans installed a full-sized basketball court and lights for Steve and his friends to practice basketball in the wintertime. This space was also used for barn dances and large parties. The children liked to watch the adult parties from the loft of the barn.

When the children were small, they spoke with their parents, grandparents, and cousins only in German. The children did not communicate in English until they went to school. Chris rebelled against speaking German at home when he got to kindergarten and found everyone speaking English. It was difficult on the first day of school, Cathy said, to be a shy little blonde girl with long pigtails and little English when other little girls had curly bobs and talked very fast. Other children were sometimes unkind to the little German children. Soon the von Ohains spoke English as often as German.

The only argument between Hans and Hanny concerning the children was at times when Hanny would say "no" to them, which Hans could not understand. Truly upset, he would ask her, "Can't you say 'yes' sometimes?" Hans did not require high standards of academic achievement for his children, but he encouraged their curiosity and challenged them with questions and many new experiences. They were allowed to seek their own goals.

The family loved to entertain, and every weekend many friends were invited to the farm from the neighborhood and from Hans's office. The lawn around the house was like a park, filled with flowers. On summer days there were chairs and tables set out for picnics. Adults could converse while the children played outdoor games and climbed up into the barns and on the roofs of the other buildings. Many photographs of the gatherings at their home are kept in family albums. Almost every Sunday the family drove to Lebanon to visit Hans's parents and Wolf and Hildegard and their family, no matter how tired Hans felt or how icy the roads.

In 1958 Hans had been away from Germany for 10 years and longed to take his wife to see the places he had enjoyed when he was younger, places she had not seen because she was young when she came to the United States. A special benefit of government civil service employment is the generous number of vacation days and the possibility of accruing a sizable number of days for a lengthy vacation. Hans had saved enough vacation days to allow Hanny and him to stay in Europe for more than three weeks. Teenager Steve, eight-year-old Chris, and five-year-old Cathy could be cared for during their absence by Hanny's father, "Opa" Schukat. Hanny's friend, Emily Walk, adored by the children because she was so much fun, stopped in every few days. Hans approached sightseeing with as much enthusiasm as he had for work. Hans and Hanny visited Venice and crossed Italy to Lake Garda, where he had spent time sailing with his brother after his college graduation. In West Germany they visited her relatives in Hamburg and went through Switzerland to France, ending in Paris. It was a relaxed time, but they missed their children.

As Albert Schukat grew older, the farm in Brookville became too demanding for him to manage. The farm was sold off bit by bit. After Hanny's mother died, Hans and Hanny decided to move into a new house in a suburb of Dayton, Washington Township. The drive to Wright–Patterson was shorter, and the family would be living nearer to Lebanon and could more easily visit Hans's parents and Wolf and Hildegard and their children, Kitty, Sylvia, and Tory.

In 1962, the family left the farm. Steve was 19, Chris was 12, Cathy was 9, and their new baby, Stephanie, was a year old. Steve, Chris, and Cathy had loved living on the farm, with endless outdoor activities in all seasons. There was space to run and a horse to ride. Chris liked being a farm boy because he could get out of school to help with the harvest. He explained, "I never really ran any of the big equipment, but it was fun to stand around and watch." Cathy cried and cried when they moved from the farm and especially missed her kitten. Soft-hearted Hans took her back to the farm a few days later to pick up the kitten, but they found it was dead, causing another flood of tears.

The children found the suburban plan of streets and yards confining, although their yard was large. The house was larger inside than the one on the farm, with more space for family activities. A city school was different for the children. When Cathy was inconsolable in missing her horse, Hans and Hanny bought her a horse and boarded it on a farm a few miles away. Hanny drove her to the farm every afternoon after school to care for the horse and ride.

Hans's stepson, Steve, graduated from high school. He left his home at the farm, with years of basketball and Elvis Presley's rock-and-roll music in the barn, to attend the University of Cincinnati and live in the dormitory. After two years of studying business, he received a draft notice at the beginning of the Vietnam conflict. He quickly volunteered for the Navy where he was eligible for officers' training. After the years of long walks and talks with his stepfather, with Hans mostly listening, Steve began to wonder, "If Papa can invent the jet, wouldn't it be fun to fly them?" He was trained as an aircraft navigator and later as a pilot based on an aircraft carrier. He did not go to Vietnam to fly because, sadly, his young wife became ill and died, leaving him with a baby, Lori Anne. The von Ohains tried to provide support during this difficult time and took their first grandchild to live with them for a few months while Steve had to be away during a training cruise. Lori later lived with her other grandparents until Steve had fulfilled his Navy obligation and could make a home for her with his new wife, Justine. They had two boys, Michael and Andrew.

Cathy continued to take piano lessons in this new location, and Hans enjoyed hearing her practice. Chris took percussion lessons from Jane Varella of the Dayton Philharmonic Orchestra and formed a rock-and-roll band with some of his school friends. They had loud practices of '60s popular music in the downstairs family room and played for a teenage social group at Normandy Church on Friday nights.

Grandfather "Opa" Schukat was happy in their new home and continued as chief cook for the family. He urged Hans and Hanny to entertain so he could cook, and he turned out big pot roasts and steaming bowls of vegetables for a full table of guests, but allowed Hanny to do her special German desserts.

Stephanie only knew the life in the suburbs, different from life on the farm known by her older siblings. Her papa was well-established in his profession at Wright–Patterson Air Force Base. Stephanie was called "the little German" at school, but prejudice against Germans left over from World War II had almost faded because of memories of the Korean and Vietnam conflicts. She had to teach the faculty and friends at school how to pronounce her name. Articles about honors and awards given to her father began appearing in the local newspapers, and occasionally someone would ask her, "Are you THAT von Ohain?"

There are many scientists who can tell stories of their friendship with Hans von Ohain during his years at the Aeronautical Research Laboratory and the Air Force Aero Propulsion Laboratory. Tony Fiore was one man who got to know him very well.

Soon after the move to the suburban home, Von Ohain telephoned Tony, his long-time friend, asking if they could carpool for the drive to Wright–Patterson. On the first day that Tony picked up Hans, he pulled his convertible with the top down into the von Ohain driveway and tooted the car's horn. When the front door opened, a giant black German shepherd dog shoved its way out of the door, raced down the lawn, leaped into the back seat of the convertible and out of the other side of the car, and took off galloping down the street. Following the dog down the street was the running figure of von Ohain, shouting for the dog to stop. When Tony stopped laughing, he helped bring the dog home.

Fiore had other stories about carpooling over the years with von Ohain. Work was often intense, and the creative process did not always end at a set time. The two scientists were able to discuss problems and new information with each other on the way home. This provided shared insights and a chance to "let off steam" away from the office. Tony remembered one day when Hans was driving. When they left the air base, Hans shifted into first and second gears, but got caught up in what he was saying and forgot to shift into high gear. The engine was roaring and Tony tried to interrupt. "Hans," he began. Hans continued with his intense explanation of whatever issue he had. "Hans," Tony said again. They traveled many miles before Tony finally shouted, "HANS, YOU ARE DRIVING IN SECOND GEAR!" and Hans shifted into high gear. Other friends of Hans said he disdained second gear and preferred to run up the car's engine in first gear until the engine was in full roar, then shift into high gear. He liked to simplify operations.

Hans enjoyed the time he spent with his family. Because the children were older and always talked about how much they enjoyed outdoor activities, the von Ohain family began annual vacation trips to upper Michigan. The family could be together for two weeks at a small cottage overlooking a lake and enjoy swimming, waterskiing, fishing, long walks, and visiting the small resort village. They relaxed and got sunburned. Hans liked to sit at the end of their dock and dangle his feet in the water, waving boaters over to him to show them the special invention he was working on, a hydrofoil ski that skimmed over the top of the water.

Sig Harder told a story from when his and von Ohain's families vacationed together in Michigan.

> One day when Hans and I were sailing and probably talking too much, the wind blew us over. We capsized and lost our rudder. He suggested that we should not buy a new one but make one ourselves. While we were shopping for materials—wood, paint, screws, epoxy (and did he love epoxy!), I asked him how he knew all those clerks and sales people. He seemed to know them because he was so nice to them. I found out that he had never met them before. Whomever Hans met he somehow touched with his heart. His example made me begin those random acts of kindness which don't cost a penny but which make us all so much happier.

Harder told more about Hans. "Hanny and I feel that every girl that ever waited on him in a restaurant fell in love with him, at least for a few minutes." And Harder says, "My mother-in-law talked time and again about a taxi ride with Hans in Munich years ago. After the cabdriver had dropped him off at his hotel, he turned to her and asked, 'Who was that gentleman? *Er war so menschlich*—He was so human.'"

In 1973, the von Ohains' son Chris graduated from Ohio University and entered the Marine Corps. His first assignment after Officers' Training School was in Okinawa, and he later was assigned to Camp Pendleton in California. When their daughter Cathy completed her master's degree in speech therapy at Bowling Green State University, she chose to take her first job in San Diego. The von Ohains missed their children, but they were happy that Chris and Cathy could be near each other in California. Stephanie was still at home attending Centerville High School. Hanny was the soprano soloist at one of Dayton's largest churches, Grace Methodist, and was active in the Dayton Music Club and other organizations. At their home, she and Hans frequently entertained military leaders from Wright–Patterson and scientists visiting from around the world, as well as other friends.

Chapter 15

Aerospace Research Laboratory

In the years when Hans von Ohain was enjoying his role as happy husband and doting father, he worked as a U.S. civil service employee at Wright–Patterson Air Force Base, first at the Aeronautical Research Laboratory (ARL) and later at the Air Force Aero Propulsion Laboratory.

ARL was formed in 1953, reorganized from the Flight Research Laboratory (FRL). Peace had returned to the United States after the tentative settlement of the Korean conflict, but tensions ran high between the superpowers, Russia and the United States, exacerbated by the Russian announcement that they were making substantial progress in building a hydrogen bomb. The United States tested an H-bomb in 1952, followed closely by the Russian announcement of the same achievement in 1953. The United States was nervous about the threat of nuclear war and appreciated that the government was performing advanced research in defensive technology.

ARL was the dominant research organization for all aspects of air technology for the Air Force for the next 22 years, until it was disbanded in 1975. Its purpose was to have a cadre of consultants, experts in all disciplines, to interface with the scientific community and advise the top management of the Air Force about the latest scientific developments. ARL carried the Air Force from the 1950s, through Vietnam in the 1960s, and into the changed political scene of the 1970s. It covered a spectrum of warfare from insurgencies to the possibility of nuclear war and a series of incidents such as the Cuban missile crisis. ARL was part of Wright Air Development Center (WADC). Other development centers were Arnold Engineering Development Center at Tullahoma and Rome Air Development Center at Rome, New York.

At its peak, ARL had 180 highly qualified and highly motivated scientists and engineers: 75 had doctoral degrees and another 75 had master of science degrees. Because of Wright–Patterson Air Force

Base as a whole and ARL as a particular asset, the Dayton, Ohio, area had more Ph.D. degrees per capita than any other city in the United States. Many scientists had attained national and international reputations and recognition through their research. Scientists were permitted to stay at ARL as long as they wanted to work, rather than to a set retirement age, and some stayed until they were in their upper 70s, still providing meaningful contributions. The laboratories recruited talented young scientists and educated and trained them to assist the top scientists. Some were civilians, top graduates of colleges across the United States. Many were members of the military who performed a tour of duty at Wright–Patterson early in their careers, before being transferred to other locations. Not many years later, when they had gained wide experience and promotions, they returned as military leaders of the laboratory. Only the top personnel were required to have security clearances because the nature of basic research meant that the scientists were far removed from actual application to Air Force projects. Scientists could come from all over the world to work at ARL, including Canada, India, Israel, Spain, and France. Appropriate rewards and recognition were given to the scientists for their work.

Staffing and budget were increased for the new organization, but in the beginning the organization's drawback was its lack of a permanent facility. It struggled in a miscellaneous assortment of substandard temporary buildings spread over Area B of Wright–Patterson Air Force Base. The limited amount of fine research equipment was housed in buildings built in the 1920s. Air Force management formulated a plan to build a new facility. Providing the laboratories with the latest equipment for current research was a second problem. In the organization's beginning, some laboratories inherited equipment from earlier installations, but most of it was outdated. To retain top competency in a subject, the scientists needed to perform research on their own. The budget was limited, and every laboratory had to be outfitted at the same time. The scientists helped themselves by looking for unused equipment or extra supplies either on base or at other installations to upgrade their own laboratories. "Midnight requisitioning" and do-it-yourself projects filled some of the essentials. In the chemistry laboratory, one scientist became a proficient glass blower to provide the exact materials needed.

Col. Leslie B. Williams from FRL was ARL's first commander. He was unhappy that, at first, the process for research at ARL was "extramural," that is, contracted out, and relied on outside sources to provide a product. He said, "We are always at the mercy of consultants and never able to make our own decisions. The Air Force needs a vigorous in-house research program with a strong, settled group of resident scientists." Under Williams's leadership, internal research programs were pursued in aeromechanics, applied mathematics, chemistry, metallurgy and ceramics, and physics. After an initial idea had been formulated, work might be contracted out for further testing.

In the propulsion area, ARL performed research in thermomechanics, hypersonics, and facilities. Development and applied technology for jet engines were the responsibility of another organization at Wright–Patterson Air Force Base, the Power Plant Laboratory. This laboratory, later named the Propulsion Laboratory, had more hands-on equipment than ARL and was involved in current solutions rather than advanced research.

ARL became a creative, exciting place to work, almost a university atmosphere. After projects were chosen, selection of individual research problems rested with the scientists themselves. A document describing the organization states, "The best judge of what specific effort to undertake is the creative individual himself: the value of the effort appears only with the passage of time." Long-range research was free of time pressure because short-range troubleshooting problems were handled by other elements of the Air Research and Development Command. Weekly meetings of the scientists allowed them to see the overall plans and problems of the organization. The sessions were noncompetitive, with discussions on how a problem could be solved best and by whom. Group discussion is a powerful tool. Today this interdisciplinary approach to pure research is an asset used in unified government laboratories in Germany and Russia, but it was lost by the Air Force with the closing of ARL. ARL's basic research was always exploring new territory, but it did not require much money because basic researchers look for the simplest solutions based on objective opinions, not on cost. The one complaint was that the programs with the highest cost and the most contractors involved were the most likely to be chosen to perform because they seemed the most important.

In all the laboratories at Wright–Patterson, including ARL, Paperclip scientists acted as mentors to young scientists. It was an incredible opportunity for the young men to be trained by these top experts of a discipline and to work with them on a daily basis. The young scientists traveled to international symposia with their mentors and met other outstanding scientists on a first-name basis. It is an impressive memory to tell that you had lunch with von Kármán or others of equal stature!

Some of these young scientists were recent college graduates. One of them, Kris Joshi, said, "I was awed to be working with these giant intellects who were renowned scientists in research." Von Ohain helped Joshi calculate efficiencies to complete one of his early projects. In a later effort, Joshi worked on multi-stage power generation, an experiment to generate power from fluid flow without any electrical power involvement. After long months of research and trials, he fired up multiple stages of the electrofluid dynamic (EFD) generator and shut off the starter power source. He was elated when it worked. Von Ohain was excited and fascinated with the success of this new technology and announced the accomplishment of this young scientist immediately to all the members of the EFD group. Von Ohain's ready praise and announcement before all the laboratory peers gave Joshi confidence that he was making a contribution to the work of the laboratory. It inspired him to work even harder. Joshi has tried to use this lesson from von Ohain in managing his own employees at the company he founded after leaving ARL.

Occasionally the Air Force complained that the work performed by ARL was not "efficient," that is, not relevant to their mission. ARL was able, again and again, to document outstanding examples of "coupling," that is, applying direct solutions to problems of the Air Force. The scientists developed many patents and presented papers in international symposia throughout the scientific world.

Von Ohain was assigned to be "supervisory aero research engineer for aerothermodynamics" in the Fluid Flow Machinery Research group of the Fluid Dynamics Group of ARL. He supervised 15 scientists and engineers. The official description of his assignment reads as follows:

> Responsible for the work of task scientists working in the field of advanced fluid flow machinery and internal flow processes. Determine internal and contractual research efforts in fields such as supersonic and transonic cascades including shock boundary layer and three dimensional flow phenomena, energy and momentum exchange phenomena by waves, viscous effects, mixing, and field interactions.

Von Ohain enjoyed selecting projects for members of his group and found choices that were interesting to the researchers and valuable to the Air Force. He liked working with the researchers to get inside a project, solve its problems, and calculate its results. He liked much less the supervisory tasks of writing reports on employees and resolving details of furniture placement and air conditioning. He was happiest away from his desk and in the thick of research, interjecting ideas and pulling the men together to discuss how they might do something. He was more outgoing than most of the Germans. He took the time to ask about the workers' families and children. He had a gentle humor that made dull projects enjoyable. He helped his coworkers and rewarded them with praise. The researchers enjoyed being his colleague. They said he was fun to work with. During these years, von Ohain kept in touch with developments in both foreign and domestic research. He reviewed and evaluated worldwide scientific reports of theoretical and experimental projects.

Von Ohain earned the attention of management by demonstrating a thorough knowledge of technology and a clear-eyed view of the goals of the organization. He was selected to provide input for a special report for the headquarters of the Air Research and Development Command (ARDC). The report, WADC Technical Note 57-13, "Survey of Propulsion Systems," coauthored by B. C. Keith and R. E. Hunter, provided an overall plan for the future of research in the field of propulsion, including a survey study of trends and research objectives in the field of energy conversion and propulsion. His work outlined conceivable propulsion systems for atmospheric and space vehicles in the present and foreseeable future. It proposed a strong program for basic and applied research. It was based on a survey among research institutes, universities, the Atomic Energy Commission, and several propulsion system contractors. Von Ohain was an authoritative and persuasive writer based on his broad classical education and the discipline of logical thought, although his grammar had to be polished from the Germanic sentence structure, and sometimes he asked, "How do you say this in English?" Upon completion of the report in 1957, von Ohain received highest praise for his outstanding analysis from the managers in the Plans and Requirements Office and from Air Force headquarters.

In the late 1950s, Dwight Eisenhower was president and Richard Nixon, vice president. Nikita Khrushchev in Russia boasted about his country's military superiority and the power of communism in the world with his statement to the United States, "History is on our side. We will bury you!" The Distant Early Warning (DEW) line was set across the northern borders of the American continent, and Americans began building bomb shelters in their backyards. Russia's launch of Sputnik in the fall of 1957 was a shock to the United States, especially when followed by Sputnik II, which sent a dog into orbit. A great sense of urgency was provoked for all research. A major program for space flight was initiated by the United States. The responsibility for manned space flight did not belong to the Air Force, but with the National Advisory Council on Aeronautics (NACA), which became the National Aeronautics and Space Administration (NASA), although the Air Force already had research and development programs for intercontinental ballistic missiles (ICBMs) and satellites. The increased emphasis and funding for space programs left aeronautical research in a secondary position. ARL looked for programs relating to space.

The Air Force now had superior research facilities for ARL at Wright–Patterson Air Force Base. Building 450, a huge complex named the Physical Sciences Building, was completed in May 1958 to house the laboratory's many functions at one location. Just after its opening, additions were planned for more space for the Fluid Dynamics Laboratory and the Physics Laboratory. Unique facilities were built: Mach 14 and Mach 20 hypersonic wind tunnels, tandem and time-of-flight mass spectrometers for ion-molecule reaction studies, high-pressure electric arcs up to several hundred atmospheres, and a high-intensity 8 MEV Van de Graaff accelerator of new design.

Von Ohain completed work on the special report for ARDC in 1957, and he resumed leadership of the group at ARL, displaying the same analytical insight, initiative, and drive he had in earlier years working on his turbojet development. His team explored propulsion and energy conversion and formulated new approaches.

Each scientist in the laboratory could call on others in completely different disciplines to have questions answered. Von Ohain's education in physics at Göttingen University had been in classic physics, not quantum physics. One day he went to Alan Garscadden, a very young scientist in the laboratory who was an expert in plasma physics, to ask him to explain lasers. Von Ohain seemed to have grasped the concept and took some books to study, but it may not have been clear. He came into Garscadden's office a few days later and asked, "Explain lasers again." Von Ohain soon was using laser technology in his research.

A major reorganization of the Wright Air Development Center in 1959 created a new organizational structure known as the System

Program Office (SPO), where the emphasis would be on procurement of aircraft and engines. The SPO would take responsibility for operational upgrades and modifications, and laboratories were left to address future concerns. Some research and development personnel at Wright–Patterson were shuffled into the new organization. SPO efforts could set predictable schedules, unlike research, and the need for research was more frequently questioned. Funding for one and then the other organization became the subject of constant high-level arguments. The laboratories tried to balance their projects among the allocations they received for what was called "6.1," basic research; "6.2," exploratory development; and "6.3," advanced development.

In 1960, ARL was administratively detached from WADC and made part of the Air Force Research Division (AFRD), ARDC. In 1961, all Air Force fundamental research in the United States was placed in one of five major research elements under the Office of Aerospace Research headquartered in Arlington, Virginia.

ARL as an organization was divided into two categories:

Physical Science Research Laboratories	Engineering Science Research Laboratories
Chemistry	Energetics
Plasma Physics	Fluid Dynamics Facilities
Applied Mathematics	Thermomechanics
General Physics	Hypersonics
Solid State Physics	Metallurgy and Ceramics

Von Ohain became the chief of the Thermomechanics Research Branch. He was assigned

> ...full responsibility for the basic and applied research program, both internal and for contracts, in the field of propulsion. Lead a team of task scientists in fields such as boundary layer and shock boundary layer interaction phenomena related to advanced supersonic flow machinery, energy transfer by viscous and wave phenomena, direct energy conversion processes between electric and fluid dynamics energy.

While this was a full-time responsibility, the Air Force, respecting his credentials, delegated him to be their "scientific leader of research in propulsion" at the same time. He was expected to step outside the responsibilities of his own office to serve as consultant to scientists and engineers throughout the Air Force, for other government agencies, and for industry. On a national and international basis, he would take an active part in coordinating committees, ad hoc committees, and other activities of the government. He traveled whenever and wherever government representation was needed.

On 1 July 1963, the lead word in the name of the organization, “Aeronautical,” was changed to “Aerospace” to reflect the expanded mission of ARL, now Aerospace Research Laboratories.

Chapter 16

Hans von Ohain's contributions

It is difficult to compile a list of Hans von Ohain's contributions to technology during his years with the Air Force. He never said what he thought were his important contributions. Much of his work was collaborative, although there are patents in his name only. Some of his work is still considered classified by the government, that is, undisclosable to the general public. A review of patents and other documentation shows that his personal contributions to research at the Aerospace Research Laboratory (ARL), and later at the Propulsion Laboratory, centered in the areas of direct energy conversion, colloid gas core reactors, thrust augmentation, and advanced supersonic flow machinery. He provided technical guidance for many other activities. A brief explanation of these disciplines follows.

Direct Energy Conversion. Most people have a conceptual or cursory understanding of the direct energy conversion of our present forms of energy, that is, coal, gas, oil, and water, into electricity. During the late 1950s and early 1960s, the technological emphasis for the Department of Defense shifted from aeronautics and aircraft to aerospace and rockets, requiring more exotic forms of direct energy conversion. Huge sums of money were directed to close the gap between Russian and United States rocket and space capabilities. New ways to generate electricity in space were investigated. Results of the investigation would also be valuable for power requirements at remote, unattended locations and under oceans.

Von Ohain's interest in exotic energy conversion concepts played well in this shifted focus. He performed and led research in areas of electrofluid-dynamic energy conversion (EFD), magnetohydro-dynamics (MHD), and magnetofluid-dynamics (MFD), along with Maurice O. Lawson, Sig Hasinger, Maj. Melvin Keller, and Ken Cramer. Research required interdisciplinary background in physics, chemistry, aerothermodynamics, and electrodynamics, including photon interaction and lasing phenomena, and used solid propellants, solar, isotope, and nuclear energy. Von Ohain was awarded several patents. ARL was well equipped with research equipment as well as the intellectual capabilities to conduct experimental investigations of these concepts.

In the EFD process, direct conversion between thermodynamic energy and electrical energy is accomplished by passing an electrically insulating working gas containing electrical charges of one polarity through an electrostatic field. In general, EFD processes are analogous to the process of a Van de Graaff generator: the mechanically driven belt that transports surface charges of one polarity corresponds to the carrier gas flow containing the electric charges also of one polarity. EFD devices offer overall compactness and operate at moderate working gas temperatures.

In the case of the conventional generator, a copper armature, which is a good conductor, is forced to move through a magnetic field inducing an electrical field on the conductor. The MHD generator uses a conducting fluid, a stream of hot ionized gas (plasma), which moves through a magnetic field. The kinetic energy of the gas stream is converted directly to electrical energy. One drawback of MHD is the requirement of an extremely hot gas stream to generate the plasma.

In the MFD process, thermodynamic energy is directly converted into electrical energy by passing an electrically conducting working gas across a magnetic field. The MFD device is analogous to an induction generator where the conductors pass through a magnetic field. Such electrically conducting gases are termed plasmas. Plasmas are composed of neutral atoms, molecules, ions, and electrons. Overall the number of ions and electrons are equal, and therefore electrical neutrality exists. Plasma as a working medium results in highly complicated mutual interaction phenomena.

Colloid Gas Core Reactor. After Sputnik, rocket propulsion became the popular topic, and jet propulsion was set aside as a "mature" technology. An electrostatic engine was investigated as a possible choice for long-distance space travel. Ionized particles accelerated to very high velocities become propellants either in an ion engine, where atoms of a single element are used, or in a colloid engine, which uses aggregations of molecules of a substance. The aggregations, or "synthetic ions," may be in the form of vapor, liquid, or solid. Colloid propulsion systems work on the principal of accelerating charged particles by electrostatic forces. Von Ohain worked with Lawson and Hasinger on a gas-cooled colloid propulsion system, which increased condensation efficiency by maximizing the number of charged particles and minimizing the neutral or uncharged particles. It resulted in a patent.

Von Ohain and William N. Jackomis worked on a colloid-core nuclear reactor rocket engine concept using fissioning uranium fuel in the form of ultrafine particles. The result was a large thrust-to-weight ratio and high efficiency. A gas was used to carry a colloidal suspension of uranium fuel directly into the propellant gas, in this case hydrogen, which generated extremely high temperatures. They described the process as something like a stream of air carrying a cloud of talcum powder. In fact, baby powder was used in experiments to check the behavior of finely powdered uranium fuel without the hazards involved with the radioactive material. Jackomis reported that the difference was that the baby powder particles are about 15 times larger than the actual fuel. "But it sure made the laboratory smell pretty," he said.

Von Ohain and Jackomis were invited to present their concept to the Atomic Energy Commission Space Nuclear Propulsion Office in 1970. Patent number 3,730,832 for "Nuclear Reactor Fuel Charging and Discharging System" was issued to Jackomis in 1973. He said, "Hans was most influential in convincing me to apply for the patent. Moreover, he was very instrumental in causing the research work to be done and in recognizing the contributing value to the field. I can say that without Hans neither the basic research work nor the patent would ever have happened. He was truly a magnificent mentor and leader."

Thrust Augmentation. Von Ohain had a long-standing fascination with vertical takeoff and landing (VTOL) aircraft and vertical short takeoff and landing (V/STOL) aircraft. He had suggested the concept for his first turbojet engine in 1935 when he took it to Ernst Heinkel. The design of a V/STOL concept (wags claim that V/STOL stands for "Very Sad Tale of Life") is difficult because, in addition to conventional flight requirements, the engine must also provide a high vertical thrust component for both takeoff and landing. Consequently, the engine thrust capability must be "matched" to the mission requirement. This usually results in a larger, more complex installation compared with a conventional aircraft. Distributing the overall weight to meet balance requirements at takeoff and landing adds additional weight and complexity to the craft. These two considerations usually resulted in V/STOL aircraft being too large, too complex, less efficient, and too expensive. Many experimental

V/TOL and V/STOL machines have been built and flown, but, to date, only the British Harrier has enjoyed any measure of success.

Von Ohain provided some very original thinking. Starting with the known properties of fluid thrust augmentation, he achieved a level of performance from a thrust augmentation device that was much higher than previous designs. His concept, which he called the "jet wing," ejects a high-pressure air stream through slots or openings on the top (suction) and bottom (pressure) surfaces of a wing over the full span of the wing. This ejected air mixed with ambient air increases the mass flow (lift) around the wing (thrust augmentation). The source of this high-pressure air is usually the engine. The increased mass flow may increase the thrust or lift by as much as a factor of two over an unaugmented system. The slots or openings, generally known as eductors, ejectors, jet pumps, or thrust augmentors, decouple the requirement for both the engine and the vertical thrust force to be close to the center of gravity. The concept solves both the engine matching and the weight distribution problems. With von Ohain's thrust augmentation device, the engine size could be adjusted to a more efficient level, and the flexibility could give an airframe designer more options for weight distribution. Von Ohain's conceptual aircraft designs applying that technology showed promise, especially compared with other V/TOL and V/STOL projects.

Von Ohain worked on the "jet wing" with young Lt. William S. Campbell of the laboratory and contributions from Roscoe Mills and Col. Robert Fontana. A paper describing their project says the process might be used to

> ...entrain and accelerate ambient air within the wing and to vector the thrust with wing flaps. V/STOL aircraft employing the "jet wing" would operate at low noise levels with little or no runway damage or surface erosion. Difficulties in transitioning from vertical to horizontal flight should be reduced. [It would offer] excellent possibilities for suppressing infrared radiation and would operate with better payload and range capabilities than a V/STOL propeller type and would have decreased vulnerability due to the elimination of complex rotating machinery.

Von Ohain saw use of the "jet wing" vertical takeoff and landing as the answer to the increasing sprawl of airports and the encroachment on farmland. His strong interest in vertical lift continued after his retirement from Wright–Patterson Air Force Base.

Unfortunately, his work reached the stage of full-scale development just as the Air Force began to rethink its need for V/STOL aircraft. Research and development funding was decreased, and emphasis moved to improvements to conventional short takeoff and landing capability. The U.S. Marine Corps became the sole advocate for aircraft with vertical capability. Marine Corps aviation acquired the Harrier, and they are acquiring a V/STOL model of the Joint Strike Fighter. Air Force units, however, remained content for many years with conventional takeoff and landing aircraft.

Advanced Supersonic Flow Machinery. Subsonic gas turbine engines are generally used in gas turbine-powered aircraft capable of supersonic speed. The supersonic air entering the aircraft inlet must be slowed to subsonic velocities before it can enter the compression system, and an inlet system with elaborate geometry was one solution for the problem. An alternative was the use of a supersonic compressor, a compression system capable of operating with supersonic air inlet velocities. Potentially, this solution had a high power-to-weight ratio. Von Ohain's investigation of the fundamental phenomena underlying the performance of the supersonic compressor focused on the blade design and resulted in his novel concept of blading with a blunt trailing edge.

Other activities. With Elmer Johnson, von Ohain studied the behavior of high-speed gases and multicomponent flow and particle separation phenomena, technically described as ultra-high enthalpy gas flows. Their work resulted in capabilities for producing ultra-high density plasmas, optical energy of extremely high photon flux, and a method of duplicating the reentry hypersonic flight environment down to an altitude of 100,000 ft, as well as novel power generation systems and weapons.

Fundamental research on swirl flows resulted in improved inertial particle separators. Von Ohain and Lawson demonstrated a device for the Navy to remove salt from sea spray. The same device provided the Army with a potential dust separator for turbine air intakes, addressing the problem that in the desert or in a dusty atmosphere a compressor ingests the sand or dust particles and the engine breaks down. Their centrifuge device collected particles in a separate filter. Von Ohain and Lawson received a patent for a device used for jeeps, trucks, and aircraft during the Desert Storm operation and on runway sweepers.

In addition to his in-house technical programs, von Ohain acted as consultant for scientists in other organizations at Wright–Patterson and for outside organizations. He presented papers at universities, NASA, and the Atomic Energy Commission. He sponsored graduate theses for students from the Air Force Institute of Technology who did practical experimentation at ARL and the Air Force Aero Propulsion Laboratory.

As had been his practice with his coworkers in Germany, von Ohain was quick to give credit for improvements to those he worked with and include names of all scientists in a study when a patent application was prepared. Surrounded by changes and inventions, he deflected attention from himself to others with the statement, "That was not invented *here!*" meaning by his own efforts.

Von Ohain not only conceived original ideas, but he had spatial visualization that gave him the unique artistic ability to clearly and accurately illustrate concepts on cross-hatched paper. His own sketches of work exist, from the first engine when he was 22 years old, to the ideas he had late in life. Charts he made for formal presentations are easily interpreted.

Chapter 17

Position as chief scientist at ARL

In 1963, Hans von Ohain received a major promotion with the U.S. government at Wright–Patterson Air Force Base. Col. Robert Fontana, commander of the Aerospace Research Laboratory (ARL), asked von Ohain to assume the position of chief scientist of ARL, replacing J. William Price, who left to take a position as executive director of the Air Force Office of Scientific Research in Washington, DC. Von Ohain protested, saying, "I am a physicist, not an engineer, and I have my own projects that I am working on." Col. Fontana replied that a physicist was exactly who he wanted in the position. For a year von Ohain was "detailed" (a temporary promotion) to be chief scientist of the ARL, but he was not relieved of his responsibilities in his group that was studying energy conversion and propulsion. In 1964, he was formally promoted to chief scientist of ARL.

After working beside outstanding scientists for many years, von Ohain became their leader. He had unquestioned technical superiority and foresight, had made outstanding contributions across many disciplines in the laboratory, and had the qualities of leadership required to derive the best efforts from the highly educated and highly skilled scientists in this unique research organization. Articulate by nature, von Ohain was most effective as liaison between the scientists and the research administration. As chief scientist he oversaw virtually all of the Air Force's in-house basic research programs in both the physical and engineering sciences. It was his responsibility to advise the commander of ARL what projects and technical changes should be studied, a responsibility made greater by the Air Force policy of assigning a new military commander as director of research for the laboratory every year or two. A commander's effectiveness depended on his understanding that managing research is very different from managing operations.

Von Ohain's official job description was as follows:

> –Responsible for advising and assisting the commander in defining and formulating a balanced, comprehensive, and coherently integrated program of basic research.
>
> –Personally conceive, plan, and carry out programs of highly creative theoretical and experimental research in the physical sciences, involving active leadership of several high-level creative scientists and supporting technicians.
>
> –Serve as consultant and advisor to the national and international scientific community in the specialized area of interest.

Von Ohain gave his perspective on the position's duties:

> The chief scientist is very influential in the laboratory. He is the right hand man. It is a staff position under the commander. The chief scientist has to know and understand the entire work which is going on: chemistry, physics, aerodynamics and so on. They chose me because they looked at my records and learned that I had studied aeronautical engineering at Göttingen, and I had my doctorate in physics and had expertise in chemistry. The chief scientist has to have a full understanding and represent the laboratory in its research work and to know what the problems were for the Air Force in these areas. Along with this qualification he has the duty, as right hand man to the commander, to tell him where the laboratory is weak and strong, where the good guys and the successes are and where it's a little bit on shaky grounds. As scientific advisor, he has to give the commander a complete picture continuously.

The men who worked at ARL gave opinions about what von Ohain contributed to the organization. "He always had so many ideas," said one. "These encompassed a wide range of programs that he couldn't possibly do by himself." He worked to develop people in the programs, very quietly giving encouragement and opportunities to allow each person to grow. Many individuals say that von Ohain was responsible for their career success. They say that von Ohain probably disliked being a "boss," but was very happy to be a counselor and teacher.

The laboratory could depend on von Ohain to carry the responsibility for the value of basic research for all the members of ARL. Von Ohain provided justification for the cost of each ARL program to local management and to Air Force headquarters in Washington,

DC. He was outstanding in his ability to communicate the details and the value of the work in appropriate language. When briefing higher authorities of the Air Force, he could summon applicable examples of how ARL research had solved a problem. It probably helped that whenever he gave a briefing, his slight German accent reminded the listeners about his impressive background, and they were very attentive.

The 1960s in the United States were the time of the long war in Vietnam and antiwar protests, of Martin Luther King and sweeping social changes, and of remarkable achievements in space. Airlines were beginning to use jet airliners, and von Ohain could review the major developments made possible by Air Force research, as well as industry. John F. Kennedy became president and brought new glamor to the White House. Americans eagerly followed his establishment of the Peace Corps and plans for an accelerated space program to have a human on the moon by the end of the decade. John Glenn orbited the Earth, followed by Russians orbiting their own cosmonauts. Kennedy faced a plunge in the stock market and the Bay of Pigs confrontation between Russia and the United States. On the international scene, the Berlin Wall was built, visibly dividing Germany into East and West, communist and free.

ARL continued basic research on all subjects and supported the war in Vietnam with a "limited war office" advisory committee with technical advice provided by various ARL disciplines. The committee was asked to troubleshoot some unfamiliar problems of guerrilla and jungle warfare. Some solutions required highly technical answers, some logical. Leonard Spialter, a veteran of research dating back to the Flight Research Laboratory in 1951, remembers some of the discussions:

> How to tell if there were people in the jungle: Find the sound spectrum of a normal jungle at certain hours of the day when there were no people, and monitor to hear the changes in bird calls or animal activity when people were present.
>
> How to identify stress fractures along rivet lines in aircraft wing structures: Record the natural frequency of an airplane when it is first built and listen for new frequencies when the wing might be cracking.
>
> A proposal for a nuclear-powered bomber which could stay up over enemy territory all the time (this was first discussed in 1951): This failed because the pilot could not be adequately protected from radiation from the exhaust that would be bounced

> back by the clouds into the cockpit. Also, a possible crash would scatter nuclear material.
>
> How to stop the Viet Cong from floating drums of kerosene and fuel oil down the Mekong River at night and hiding them in the daytime: Booby-trap a few drums and allow them to go to the storage area to blow up.

Retired ARL scientists say that they worked together like family, knowing each other's capabilities. There were theorists and experimentalists on every subject. Scientific discoveries were made by a scientist huddling over a desk working calculations on paper or by a tinkerer attaching a new part to an old apparatus, but former employees tell that there was another way to get the thoughts flowing: walking the halls. They say it was not unusual to find scientists pacing the long hallways for long periods, absorbed in their thoughts and unresponsive to any greeting.

The laboratory had its own computer, a Burroughs E-101 used by the Mathematics Research Group since 1951, replaced by an IBM 1620 in 1966. "For my own use, I learned programming," Spialter said. "Soon I was teaching the Ph.D.s and others in the lab FORTRAN, machine language, and symbolic language." The scientists had to keep up with their peers in academe and industry.

The Computer Center of Wright–Patterson Air Force Base had large commercial machines including the ERA 1103 in 1955, the Univac Scientific 1103-A in 1959, the IBM 7090/7094 in 1961, the CDC 6600 in 1970, and the CRAY X-MP in 1987, but for each the use was tightly controlled. An organization did not have to pay for use of computer time, but it was allocated a fair share of time with others on the base according to an algorithm. In 1972, at the time of the CDC 6600 after the installation of the NOS/BE operating system, the staff of the Applied Mathematics Laboratory considered this to be a delicious challenge. Paul Nikolai, who was there at the time, smiled when he reported, "Our private solution of the algorithm resulted in ARL being able to use a large percentage of all computer time for the entire base."

Von Ohain never learned to use computers but respected them as a tool. "We believe that almost everything is possible with the computer," he said. "That is correct as long as we're talking about optimizing existing concepts. But when we are talking about creativity, about new ideas, then there is no better computer than the human brain!"

Scientists were able to work directly with von Ohain, the chief scientist, to verify their work or look for additional direction. The sessions often took place with exchanges taking the form of two to three hours of blackboard calculations for facts and details. Von Ohain would say, at the end of the session, "Don't erase that; be sure to capture that and go over it again." The information was copied off the board by the scientist to add to the project folder.

Von Ohain was a great diplomat. He never told anyone that they were wrong or had made a mistake. His closest coworkers figured out, after long association, that if von Ohain did not agree with some conclusion he would say, "I do not understand." The natural reaction would be for the presenter to rethink what he had done, often arriving at a different answer. Ken Cramer told of a briefing he made to von Ohain on a particular effort dealing with arc-heated gas. Cramer was very certain of his facts and froze when von Ohain said, "I don't understand." When von Ohain saw the look that passed over Cramer's face, a slow realization came to him that his coworkers had seen through his method. Von Ohain started to laugh. "I didn't mean that," he said. "I really don't understand!"

ARL scientists came from many countries. They traveled together with interesting results. One rainy night the German Hans von Ohain and the Scottish Alan Garscadden found themselves lost driving around Boston looking for their hotel. They stopped at a gas station and received good directions, but the attendant was curious about them, men with two thick, distinctly different accents. "Where are you from?" he asked. "Ohio," they answered. He puzzled for a moment and observed, "You don't sound like you are from Ohio."

There were many business trips (temporary duty or TDY) out of town for the scientists, either acting in advisory roles and presenting papers or for attending large conferences. The $25 per diem allowance at that time was barely sufficient to cover hotel room, food, and car rental. On one trip, four of the top management of ARL, von Ohain, Tony Fiore, Elmer Johnson, and Ed Callan, were traveling together to an AIAA meeting in Pasadena, California. Callan had a brilliant idea. At a junkyard he bought a car for $25 that barely ran, and they all shared in the cost. They planned to return it at the end of the week and get their money back. Callan said, "After all, we're all top-flight engineers. We can keep the car running." They drove from the hotel to meetings and out to dinner in the evenings with little trouble, just a few incidents when someone would have to climb out of the car to adjust or reconnect something. One evening at a restaurant with valet parking, the dilapidated hulk sat in the driveway in front of the restaurant for two hours because the valet could not get it started. Everything went well until the last day when they were on their way to return the car. The plan was to get a taxi from the junkyard to the airport for their return flight to Dayton, but the car died on a California freeway. After a conference among the men, Callan pulled the car into the grass beside the freeway. They threw their suitcases over the 6-ft fence beside the highway and climbed over. Johnson was the oldest and had the most trouble getting over. They picked up their suitcases, walked to the nearest gas station, and telephoned for a taxi to the airport. They left the car sitting beside the highway in California.

The Air Force encouraged the chief scientists of the several disciplines to keep in communication with each other. Von Ohain at ARL traveled many times with the chief scientist of the Air Force Aero Propulsion Laboratory, Bill Heiser. For one conference, the men took their wives and the Heisers took their seven-year-old daughter, Katie. The Heisers and Hanny went off together and left Hans to babysit the little girl. Whey they returned, she was all smiles. Hans, who loved children, had entertained her all afternoon with stories of his work. The Heisers reported that on the way home the next day, the pilot asked the little girl to come up front and see the cockpit. Katie startled the pilot by telling him, "Yesterday I spent the afternoon with the man who invented the jet engine." Of course, the pilot did not believe her.

Marv Schmidt tells of the time that he and Jack Richens traveled with Hans to Tullahoma. Government quarters have different levels of accommodations according to the rank of the person, and Hans, with his formal German background, expected to have a very nice room. The desk clerk was young and did not respond when Schmidt said, "He needs a special room because he is a P.L 313." (Hans had been appointed to his high position under Public Law # 313.) When the clerk asked what this compared with in military rank, Hans was casually leaning against the desk with his back to those making arrangements. They heard him mutter under his breath, "Brigadier general." He got a suite.

Levity was in short supply in the studious atmosphere of the laboratories, but it did not prevent some pranks. One autumn the big round vacuum spheres on the roof of Building 450 were painted with a primer coat that was almost orange in color. On Halloween night they were decorated with jack-o'-lantern eyes and a toothy smile. The action was blamed on the youngest lieutenants in the office, but no one confessed and no one was punished.

Hans liked funny stories and American slang. He was quick with a teasing remark, which was always kind but had a refreshing perspective. He liked to tell this "underground" joke from Germany:

> In a German elementary school geography class, the teacher was explaining locations on the globe. She asked, "Who can point out Germany on the globe?"
>
> Little Fritz was first to volunteer. He rushed to the front of the class, spun the globe round and round, and put his finger on the largest land mass. "This is Germany!"
>
> The teacher corrected him. "No, Fritz, that is Russia. Try again." Fritz spun the globe several more times, then stopped it and put his finger on another large land mass. "This is it!" Again the teacher corrected him. "Fritz, that is Africa. Look again and see if you can find Germany." Fritz spun the globe furiously, then stopped it and pointed with assurance. "This," he said, "is Germany." The teacher said, "That is the United States."
>
> The teacher slowly turned the globe and pointed to Germany, in the center of Europe. Little Fritz stood looking at it for a moment. Hesitantly he said, "Does the Führer know this?"

In 1970 von Ohain was selected to attend the Federal Executive Institute, a six-week resident school for executive education in Charlottesville, Virginia. Sensitivity training was a trend at the time: loosening up to listen for new ideas; brainstorming, that is, listening to every idea, no matter how wild, and giving it credence; and showing respect for each person's contribution. All of these techniques describe how Hans von Ohain already worked and were inherent to the successful pursuit of basic research. Seeking consensus in opinions would have been harder for him. For those who knew this dignified man, visualizing his participation in some of the clever exercises of the training takes a long stretch of imagination; for example, the exercise when all the participants were required to lie on their backs on the floor, swing their arms, and shout motivational slogans. He enjoyed amusing friends at his home with this experience from the training.

Because the men were away from their families for a long period, the wives and children were invited to Charlottesville for a long weekend and were included in group sessions. In a session for the wives, each woman was asked, "If you could change one thing about your husband, what would that be?" While answers given by most wives included a list of bad habits, Hanny von Ohain answered demurely, "I just wish he wasn't away from home so often."

The span of a long lifetime brings many changes. Hans von Ohain had been born in a formal, conservative society in Germany and educated in the strictest discipline. He lived through a bitter war when his country was on the losing side and successfully adjusted to a new culture and new language when he came to the United States.

The 1950s in the United States were times of a simple, peaceful life dedicated to home and family, in spite of the interruption of lives cause by the Korean conflict (specifically not called a war).

In the 1960s, social revolution began. Protests were staged against racial segregation and the government's Vietnam policies. Women demanded equal pay for equal work. There were new words in the vocabulary: blastoff, orbit, Minuteman, Polaris, Gemini, Apollo. Von Ohain, who as a child had seen humans fly in wooden airplanes with fabric-wrapped wings, saw a man walk on the moon in 1969.

The 1970s saw veterans return home without glory and without jobs after their country lost its first war, the longest in its history, the Vietnam campaign. International power struggles led to an oil embargo and a gasoline shortage. An American president resigned in disgrace. Jimmy Carter, our gentle President, championed human rights while high unemployment, double-digit inflation at home, and a devalued dollar abroad concerned Americans about their position in the international scene. And von Ohain knew that the SR-71 Blackbird flew from New York to London in one hour, 55 minutes, and Viking I made a soft landing on Mars.

Chapter 18

Air Force Aero Propulsion Laboratory

In 1975, von Ohain had been chief scientist of the Aerospace Research Laboratory (ARL) for 11 years. The pendulum of government priorities swung away from basic research, and no one stepped in to save it. In spite of cries from many scientific circles nationwide, ARL was disbanded.

The rationale for the change was never clear. The apparent explanation was that because the U.S. Air Force's Scientific Advisory Board (SAB) was made up of academics and defense contractors, and either faction, looking to their own interests, may have stated that pure research was inappropriate for the government to pursue or too expensive for the Air Force to continue. The SAB correctly identified major abuses of research money for nonresearch purposes in several Air Force locations, but none were identified at ARL. The SAB recommended that research funding should be made available for external investigators, and the Air Force decided that much of it would come from ARL. The decision was made that research would be performed by the Air Force only when a customer was willing to pay for it. This put ARL out of business. Direction was given that engineers employed at the laboratory should be absorbed into other jobs with more practical applications, and physicists and chemists should be released to go to universities. Many scientists ended up in unsuitable positions.

Closing ARL cast a long shadow over technical innovation in the United States for years to follow. Protesters said that the Air Force was losing its greatest asset for continuation of international superiority in technology. Apparently there was no turning back. Elmer Johnson, Roscoe Mills, and other outstanding leaders that had put the United States in the forefront of research since the late 1940s left the Air Force.

ARL closed and, after a substantial layoff of personnel, the remaining ARL scientists were absorbed into the Materials, Avionics, Flight

Dynamics, and Aero Propulsion Laboratories. Von Ohain said, "The Air Force felt it would be more meaningful to get the research directly in touch with the development laboratories." Von Ohain was selected to be chief scientist for the Air Force Aero Propulsion Laboratory (AFAPL), replacing Bill Heiser, who went to Arnold Engineering Development Center (AEDC) at Tullahoma. Von Ohain explained, "Since my background was in gas turbines, propulsion, energy conversion, physics, and chemistry, they said, 'We don't want someone from the outside to take over.' I was almost ready to leave the Air Force at that time to go to industry, but they needed me. I thought, 'I will do it for a few years.'"

Von Ohain did not end up at AFAPL by chance. Bill Heiser was well acquainted with von Ohain's desire to "work one more time in airbreathing propulsion," as well as with the painful uncertainties of "disestablishment." The chief scientist position at AEDC was coming open at the same time, and Heiser felt that his abilities were needed there. He agreed to transfer to AEDC provided that von Ohain replaced him at AFAPL. This was an extremely successful arrangement because it put von Ohain in the right place and allowed them to work together for several more years.

After his illustrious career in Germany, von Ohain chose to remain with Air Force research for an entire second career in the United States. He might have earned far more money had he gone with industry. When asked about his decision, he said that he liked what he was doing and he felt that the salary provided by the government was adequate for a good life. The Air Force research community gave him an opportunity to work on the leading edge of technology. Von Ohain appreciated that the ideas developed in his Air Force office could be followed by real testing and development, which followed his personal pattern of having an idea and trying it out. With the Air Force, incredible ideas could be examined—his own, and, with his encouragement and leadership, those of the fellow scientists in his organization. The prestige of his position with the United States government allowed him to stay in touch with other international figures in his discipline. Practically speaking, von Ohain's decision to stay in Dayton was based also on the situation established early: his wife's family was in Dayton and his parents and his brother's family lived in nearby Lebanon, Ohio.

The Air Force gained from his decision to continue to work with them. The technical knowledge available to the United States was increased by von Ohain's generous policy of helping and advising others throughout the scientific world and, in turn, receiving information from them.

The mission of AFAPL was to make improvements and innovations for turbojet propulsion based on programs already in existence. Von Ohain's responsibility increased because the size of the organization was greater, from the approximately 250 employees of ARL to almost 450 in AFAPL. There was increased funding provided by the national budget. Von Ohain was expected to:

> –Evaluate total laboratory technical program to determine its adequacy, fruitfulness [sic] and emphasis.
>
> –Identify research gaps and analyze advancements in a variety of technical fields to determine their influence on laboratory programs and objectives.
>
> –Develop methods of updating in-house competence and recognizing engineers and scientists to increase technical excellence of the work and to motivate the professional staff.
>
> –Serve as advisor to the professional staff of the laboratory and to other organizations of all echelons concerning the laboratory mission and program.

Laboratory commander Col. James Nelson wanted to find out what in-house projects were in process throughout the laboratory at that time. He hired Mark Dunham, the recently retired deputy director of the Propulsion Laboratory, to identify all the projects, assess the value of proceeding with them, and determine the cost that would be involved. He was to work with von Ohain, then chief scientist, in evaluating the total effort, project by project. From every individual, they found a number of creative ideas that could be pursued. After two months, they reported to Nelson, "You have no idea all the stuff that is going on!" The initial look was followed, over the next several months, by a detailed analysis of each of the laboratory's operating divisions. The results were incorporated into the annual budget and planning for the entire laboratory. Von Ohain, with this total view of the in-house efforts, was able to ensure that each project was valuable, not only to the individual scientist/engineer involved, but also to the total laboratory program.

Recruiting personnel for the AFAPL was no problem. The Air Force had been wise to put Hans von Ohain in a prominent position. His name and reputation were becoming well known. A rush of top graduates from universities vied for the opportunity to begin their careers working with von Ohain.

Von Ohain accepted the new position with grace and enthusiasm. He said:

> It was the same type of activity [as ARL] with an exception: in addition to consulting with the commander I had the entire in-house program of this big laboratory under my direct responsibility. That means I really had to work with all the people who did in-house work, and I had to know exactly what they were doing, and also be able to tell them when directions were to be changed. Of course, I did it in a courteous way because the head man [supervisor] was the division chief. It was a delicate matter for a staff man of the commander to tell [one of the workers] what to do, what not to do, how to do it, when what he did wasn't very good. I had to give him constructive criticism and also inform his direct boss. That poor guy had two bosses: one, the chief scientist and one, his division chief. It was very clear that was my responsibility, and I handled it in such a manner that problems did not arise by being very open to the line organization people who were actually the bosses of that man.

To von Ohain's credit he recognized the stature of the scientists who worked at AFAPL. He nurtured each one and maintained the laboratory's reputation for innovations and creative solutions for Air Force projects. Being a basic researcher was somewhat discouraging for rewards because there was no technical "ladder" for promotion. There was only one position to be promoted into, chief scientist, while the promotion ladders for the contract managers included steps to branch chief and division chief. Von Ohain made efforts to initiate monetary awards for special achievements by scientists.

Von Ohain worked closely with Cliff Simpson, who in 1945 had interviewed him at Kolbermohr, Germany, and was, in 1975, chief of the Turbine Engine Division of the AFAPL. The two turbojet experts had continued their early friendship during the years when von Ohain was in ARL and Simpson was in the Power Plant Laboratory, later named Aero Propulsion Laboratory. Now that they were in the same organization, von Ohain explained the difference in their responsibilities: "I was chief scientist, staff advisor to the commander, and Cliff Simpson was in the line organization. We [both] wanted to be informed about directions in order to judge whether the in-house program should be strengthened."

Simpson was an early U.S. pioneer on the turbojet concept beginning in the early 1940s with the Whittle engine and had developed a legendary reputation concerning turbojet development. His advice was accepted at the highest levels of the government. Simpson was a colorful character, and many stories are told about how he loved acting like a Kentucky country boy to obfuscate or gain time when he was not yet ready to make a decision. Simpson had weathered the early 1960s when some of the highest Air Force management in Washington, DC, had been promoting the opinion that the turbine engine was a mature device that needed no further funding. Von Ohain commented:

> Cliff had a big problem with administrators who thought the gas turbine was no longer really a research item, but just absolutely a straightforward, innocent development item. It was a totally foolish time period when they said "There's nothing in gas turbine development any more. There's only incremental improvement to be made." They called it "nut polishing." Nut polishing! There were a heck of a lot of nuts that had to be polished. Aero propulsion was critical and needed far more research to create lighter and lighter engines, more economical engines, more powerful engines, higher speeds, and so on. They were wrong, but it was a general trend because the sexy thing was the rocket. Everything was focusing on making rockets—large rockets, space rockets, intercontinental ballistic missiles.

In 1965, Simpson had succeeded in convincing the highest Air Force and Defense Department echelons of the significance of a major technological breakthrough, the high-bypass turbofan engine. The high-bypass engine had a role in the establishment of propulsion systems of highest possible thermodynamic and propulsive efficiencies, that is, very high compressor pressure ratios, very high turbine inlet temperatures, and very high bypass ratios.

From AFAPL, Simpson and von Ohain traveled frequently to Washington, DC, as advisors to Air Force leaders. Simpson was the expert for practical solutions, and von Ohain, for explanations of the more esoteric technical concepts. Both knew the key people. Both were articulate and persuasive. Both kept the Air Force on the best course for achieving its mission.

Von Ohain described Cliff Simpson:

> A first-class, hundred percent excellent division chief. I was amazed at his grasp, technically and scientifically, and his ability to pick out the right thing. Cliff was much better in the managerial thing than I because his decision was [brought hand down on table with a bang] "That's it!" I envied Cliff often for that. I could see too many transitional things. Sometimes I'd say, "Cliff, this sounds a little bit too black and white to me, with things being either bad or good. I see a little bit more gray area in between." I saw so many ifs and buts. Cliff came to a clear decision, and I have never seen, in the time period that I knew him, that he had done the wrong thing.

Simpson and von Ohain joked with each other, friendly disagreements where one would point a finger at the other and say, "You are a physicist! What do you know of practical engineering?" or the opposite, "You ignore science!" Discussions between the two in the hallways should have been recorded. The coworkers who observed the interchanges would have loved to hear what these two world-renowned experts were deciding.

Simpson said of von Ohain:

> I developed a deep respect for this man's character, particularly his kindness, understanding, and humble demeanor. There have been many discussions, usually rounding out and defining an engineering problem like "Does the J65 have one compressor stage too many?" or less technical areas like "In C.B. radio, what is a 'beaver?'" My great respect is based on Hans's cleverness and the determination it took to be first in a major invention or development. The pioneer is faced by doubts, failures and uncertainties that assail the mind, numb the senses and urge you to quit over and over, when it would be so easy to do so. Hans has never quit or given up, and those of us in the turbine engine business now begin by knowing that with just sweat and work, technical success will follow. Thank you for your wise counsel.

In those days, the reality was that the Army and Navy had little capability for propulsion development compared with that within the Air Force. The Army had a small engine technology program,

Small Turbine Advanced Gas Generator (STAGG), which closely coordinated with AFAPL's programs for Advanced Turbine Engine Gas Generator (ATEGG) and Advanced Propulsion Systems Integration (APSI). The Navy had been directed by the Defense Department-level office overseeing these programs to integrate their small, separate programs into ATEGG and APSI before funding would be released. This became the Joint Technology Demonstrator Engine (JTDE), funded and managed jointly by AFAPL and the Navy technology office at Trenton, New Jersey. NASA Lewis (now Glenn) Research Center was returning to airbreathing propulsion during this period, and ever-increasing review and coordination was necessary between NASA Lewis and AFAPL to avoid duplication of efforts.

At this time, the top leaders of engine companies, including William Brown of Pratt & Whitney and Gerhard Neumann of General Electric, could be found in Simpson's office for appraisal of their major technology plans and projects. They knew that they would receive a frank opinion of their plans, and suggestions for how those plans could be of more interest to the Air Force, while applicable to both the military and commercial interests of their respective companies. Von Ohain often participated in these sessions, either directly or in discussions with Simpson later. Von Ohain's technical expertise and insight were invaluable. In turn, he was able to become much more familiar with the overall capabilities and business strategies of the propulsion companies, areas where he had not had a great deal of opportunity or activity while at ARL.

Von Ohain kept in contact with the many activities in process in the laboratory. He remembered, "I discovered some things which I considered to be utterly crazy and impossible, dumb and crazy, because they wanted to save money in ways that were impossible to do." Von Ohain's management style, walking the halls and looking in on people, saved the Air Force from more than one costly wrong decision. "I learned the benefit of what can happen in an organization when a guy can put his hands in his pockets and walk around to see what people are doing, and ask questions," he said.

He related one example:

> I asked a few outstanding young Ph.D. engineers what they were doing and what the idea behind it was, and they told me with pleasure. I said, "Well, what you have proved is that the [wind] tunnel cannot be built that way. Are you sure?" "Yes," they said, "We are sure but it's very tough to get that message

> up the line." Facilities research work was, of course, my cup of tea. I said, "Well, now, that's no problem for me. On the contrary, that's kind of an amusing operation, a pleasure. I can tell it as a tough management representative." I told the vice-commander. After I had made some friendly but devastating comments, he countered, "We have a great consultant." I said, "That consultant is technically and scientifically first class, but don't forget, a consultant has no right to say 'it cannot work' unless you ask him. Did you ever ask the question? Let me ask the consultant in your presence what he thinks about it but be prepared that his answer could be devastating." A number of horrible, bad things came out. They were so happy that a fresh wind was blowing in the laboratory. The working people were happy that I was there and could give credit to the guys who actually knew about problems but couldn't penetrate the layers to come to the top. Things were now moving right, fixed with a little bit of money.

Von Ohain's professional life was a series of meetings for presentations about projects in the lab. In spite of many years away from German discipline, he kept his precision concerning being on time for meetings. When the appointed hour for a meeting was reached, all personnel were expected to be in their places. Those that slipped into his meetings late were greeted with a nod from the chief scientist and his words, "I forgive you." Most were not late a second time. Von Ohain can be pardoned for instances when he heard a subject discussed at length for the hundredth time, for him, and he closed his eyes to analyze more deeply or to remember the first time he had heard such information. Just when everyone was concerned that he had been asleep, he would lift his head and ask probing questions or make incisive decisions.

Von Ohain's work with the Air Force and in international circles drew more attention to his early accomplishments in Germany and the first flight of the turbojet engine. Honors began to be given and continued the rest of his life. An award was generally given at a formal dinner followed by speeches by leaders of industry, government, and professional aeronautical and engineering organizations. There were perks of limousines and fine hotels as well as airline arrangements. One honorary dinner included a cruise. Hans's obligations were to dress up, smile, and sign autographs. He was concerned that he had offended Ernst Heinkel's son at the time they first met because he spent his time signing autographs (and missing dinner) instead of talking with the younger Heinkel. Von Ohain accepted invitations with grace. With modest humor, he would say, "Hanny and I will be going to Washington [or New York or San Diego or Munich] because Hanny likes to go to parties and eat good food and meet nice people." He depended on Hanny as she did him. They totally trusted, respected, and accepted each other. He appreciated all she did to make his life pleasant.

In 1978, Col. Philippe Bouchard, at that time the commander of AFAPL, invited Sir Frank Whittle, then a member of the faculty of the U.S. Naval Academy, to come to Dayton. Whittle would join von Ohain, chief scientist of AFAPL, for meetings at the air base for two days. One day would be at the laboratory and one at the Air Force Museum, for public presentations together as part of the Air Force's commemoration of the 75th anniversary of the Wright brothers' first powered flight. Whittle and von Ohain had met in 1966 at the Goddard Award, but they had never answered questions in an open forum.

On the first day, a turn-away crowd of scientists at the laboratory was given the opportunity to brief their projects to Whittle and von Ohain and ask the two jet engine inventors questions. It became a "free-wheeling give-and-take conference" and could have continued several hours longer because the scientists discovered the two inventors had faced the same problems, technical and bureaucratic, that were still commonplace in 1978. The open session at the Air Force Museum brought forth an enormous crowd from all over the United States. Among the guests were three close associates of von Ohain from Germany now employed in the United States: Hans Mauch of Mauch Laboratories, Helmut Schelp of Garrett Corporation, and Max Bentele of AVCO Lycoming.

The encounter was described to have "energy, spontaneity, and excitement." The proceedings of both sessions make fascinating reading. They were recorded in an Air Force Systems Command Historical Publication, *An Encounter Between The Jet Engine Inventors, Sir Frank Whittle and Dr. Hans von Ohain*. The two compared notes and gave their own reminiscences of difficulties creating engines. It was kept light and friendly because of the high regard each felt for the other.

Whittle set the light-hearted tone of the sessions by joking about the aircraft "with a hole at each end." He told a story that took place in 1942 when a British Wellington aircraft was being used as a flying test bed. It had two Bristol piston engines and propellers as its normal power plant and a W-2B engine at the back where the rear gun turret had been taken out. As the pilot approached a B-17, just for the fun of it, he cut both his piston engines and feathered his propellers, opened up the W-2B, and overtook the B-17. The crew of the B-17 thought they saw a Wellington passing them with both engines failed.

Whittle and von Ohain talked of engine durability. Whittle said that the British Air Ministry would not allow the jet engine into production until it had passed a 125-hour type test, which his did in 1943, while the German Air Ministry was letting them go at 25 hours. Von Ohain countered, "The reason was that in the wartime a normal [German] fighter had not more than 10 hours total lifetime."

As a pilot, Whittle wanted to fly his own invention, but the government would not give him permission. They said he was too valuable, and they did not want the inventor to fly the invention. He had to cheat to do so. One day at his airfield he said, "I want to taxi the Meteor prototype," and there was no one to stop him because he was an air commodore. "So I got in it and I taxied it and taxied it. And do you know, it took off!" He said he was not used to speeds of more than 200 mph, and he flew very carefully. He said, "It was about the nicest aeroplane to handle that I've ever experienced." When he returned, one of the engineers commented, "I watched you. I didn't think the Meteor could fly so slowly." Von Ohain, who was not a pilot, quoted Gen. Galland's description of flying Germany's Me 262 jet aircraft: "It is as if angels push you through the air!"

After discussions about technical and bureaucratic problems they had overcome, the two inventors joked about "the four-pound chicken test." Birds ingested into the engine were a problem even in the beginning of jet propulsion. Fan blades were tested for durability on the test stand by throwing a four-pound chicken into them to see if they could withstand bird strikes.

After the formal session was over, Bouchard asked the two inventors the questions everyone wanted to know. Has turbine engine technology run its course? Are we just working the edges with no major improvements to be expected? The two men rolled their eyes to each other, and after a moment's conference, they smiled and gave their reply. "If the government were to give us $100,000 today and stay out of the way (that is, remove the paperwork and restraints), we could show you all sorts of potentially significant improvements!"

Tom Curran, who succeeded von Ohain as chief scientist at AFAPL, told a memorable incident from those days. When Whittle was in town for the two-day sessions, Curran and several others took Whittle and von Ohain for lunch at Stouffers Hotel. After lunch, as they turned onto the highway to return to the base, a truck ran a red light and missed, by mere inches, crashing into the side of the station wagon next to where Whittle and von Ohain were sitting. After all the passengers of the car recovered a normal heartbeat, they commented that Dayton, Ohio, came very close to making national and international news. They could see the headlines: "Both Inventors of the Jet Engine Killed in One Accident in the Home of Aviation."

Hanny and Hans invited Sir Frank, as he was called, to come to their house for dinner while he was in Dayton. Hanny remembers that Sir Frank looked very frail. Hans and Sir Frank were dear friends after years of their continued contacts. Hans was concerned when his guest ate very little. Sir Frank explained that he had been in poor health for several years. Von Ohain had robust good health and continued his daily walks with the family's German shepherd dog.

In 1978, the Air Force Systems Command (AFSC) wanted to repeat an initiative similar to the von Kármán report in the late 1940s, this time to be called *Look Forward 20 Years*. Von Ohain and his assistant, Jim Petty, gathered the more inventive laboratory researchers into meetings for "blue skies" brainstorming of the future. The goal was to envision developments the laboratory might make in 20 years in order for management to work backward to see how the progress could be made to happen. Jim Petty remembers that two of the more far-out ideas were anti-matter propulsion and metallic hydrogen as rocket fuel. *Scientific American* had an article on the latter in late 1999. Petty said, "Maybe we weren't so crazy!" A report was prepared and von Ohain presented it to Bernard Kulp, chief scientist of AFSC.

Von Ohain had reached retirement age and had the security of a government pension for his everyday needs. He had many ideas he wanted to pursue beyond the limitations of the AFAPL. He said that he had an idea that would revolutionize the Air Force and that he did not want to delay working on it any longer. He announced that he would retire in January 1979.

A surge of enthusiasm by his peers for von Ohain's contributions during the years in Germany as well as during his Air Force career led to the compiling of a commemorative volume, *To Honour Hans von Ohain,* presented to von Ohain at his retirement dinner. It included a lengthy biography by Hermann Viets, *Hans von Ohain—Contributions of Excellence,* based on stories von Ohain had told and official records. Many who had worked with von Ohain at the laboratory contributed a short description of what they had done together.

Von Ohain and his brother, Wolf, were called early in the morning on the day of his retirement dinner to come to the nursing home in Waynesville, Ohio. Their mother, the beautiful Kitty, had died during the night. At the Officers' Club at Wright–Patterson Air Force Base that evening the speeches and the honors of the formal dinner could not change the underlying mood of sadness for von Ohain. Sometimes life's problems interrupt the most wonderful occasions.

Von Ohain was described on that day as a "bluff, hearty, jolly man, looking 20 years younger than his age, and fascinated not half so much by the past as by the future." He accepted the many tributes with his usual modesty and looked forward to the rest of his life.

Base housing/barracks, 1947.

Hans von Ohain's indentification picture at Wright–Patterson Air Force Base, age 35.

Hans von Ohain in 1949.

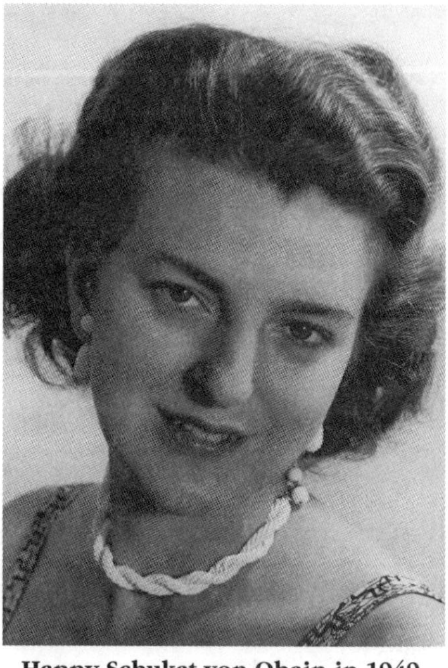

Hanny Schukat von Ohain in 1949.

Aerial view of the Schukat farm, 1955. The house is at the left, hidden in a grove of trees. Above it is the chicken house. Next are the bank barn and cattle barn.

Family and friends gathering at the Schukat residence, 1949. Pictured, left to right, are Wolf Pabst von Ohain, Hans's father; Dick Becker Jr. and his wife; Kitty Pabst von Ohain, Hans's mother; Albert Schukat, Hanny's father; Helena Schukat, Hanny's mother; Erna Becker; Hanny Schukat von Ohain and Hans von Ohain; Hildegard and Wolf von Pabst, Hans's sister-in-law and brother; and Richard Becker Sr. Erna and Richard Becker had come to the United States earlier and encouraged their good friends, the Schukats, to emigrate to the United States in 1929. The Schukats lived with the Beckers until they were able to establish their own home in Dayton.

Hans von Ohain as chief scientist for the Aeronautical Research Laboratory (ARL).

Cartoon for Hans's birthday in 1962. This refers to the laboratory's use of baby powder to test the colloid-core nuclear reactor.

The von Ohains moved to this home in Centerville, Ohio, in 1962.

Hans with Cliff Simpson. The turbojet experts worked side-by-side at the Air Force Aero Propulsion Laboratory (AFAPL).

Hans reviewing his notes before making a speech.

Hans receiving the "Web of Space" award from John Glenn, presented by the National Air and Space Museum.

Sir Frank Whittle with Hanny and Hans von Ohain photograhed at the National Air and Space Museum.

Helmut Schelp, Sir Frank Whittle, Hans von Ohain, and Max Bentele at Wright–Patterson Air Force Base, 1978.

Hans lecturing to a class at the University of Dayton, 1981.

Hans von Ohain photographed in front of a modern turbofan intake.

Hans working on the human-powered watercraft in his workshop in the garage of his Florida home, mid-1990s.

Hans posing for a picture of "flight" with his daughter Stephanie beside him.

Chapter 19

Work after retirement

Von Ohain said that retirement would be a time for him to think his own thoughts. At first, he worked on some projects independently at home, in his own workshop and office. As his daughter Cathy said, "His hands were always making or creating something at home. He worked on a hydrofoil ski and discussed with my brother Chris the possibility of designing and developing a human-powered watercraft." She added, "But he didn't fix plumbing."

Very soon his scientific interests spurred him to find formal employment. John Minardi, who had been a contractor for projects at the Aerospace Research Laboratory (ARL), and Mo Lawson, von Ohain's old friend, invited him to work at the University of Dayton Research Institute (UDRI). Dale Whitford was their supervisor. The organization already employed some prominent scientists, and von Ohain believed this group of engineers and scientists could help him further his personal projects without the limitations he had felt in the Air Force. In addition, UDRI gave generous financial rewards for inventions developed under their auspices. Soon von Ohain brought his friend Frank Wattendorf to work with the group on projects. Von Ohain worked several days a week at UDRI for almost 10 years.

In dazzling brainstorming sessions, these top scientists at UDRI expanded on ideas they may have first conceived in their ARL and the Air Force Aero Propulsion Laboratory (AFAPL) years and prepared proposals for industry and the Air Force. Interaction among these peers was exciting. Many of their projects were concepts on paper, and some required hardware to be built. Because of the outstanding credentials of von Ohain and other scientists at UDRI, members of the group were requested by professional organizations and the Air Force to write and present a number of papers for symposia.

Being a part of a contractor's office was different from government work. Ideas had to be developed and polished for presentation to industry or the Air Force to receive a contract to proceed further. The concepts had to be tailored to the needs of a particular user. Von Ohain had the talent and experience to represent UDRI and his own ideas, but he had to accept that what the team considered to be an outstanding offering might be rejected because it might be too costly or did not seem feasible to those listening. Often the small exploratory contracts of $100,000 or less that they received could not support sufficient development to prove an idea valuable. Von Ohain's projects were an internal bypass monorotor engine (IBME) concept, a swirl-flow heat exchanger, and aerodynamic confinement of small particles in a particle- (colloid) fueled nuclear reactor. He received more patents.

Von Ohain's curiosity about how things worked continued during his time at UDRI. John Minardi showed him a new piece of equipment, a smoke-generating machine that would assist in their flow visualization tests. Von Ohain was eager to try it, forgetting that the university had recently installed a new smoke alarm system. He turned it on and, within minutes, managed to clear all personnel from the four-story office building.

The University of Dayton soon asked von Ohain to teach graduate classes in propulsion and thermodynamics. The university was a leading school of engineering and advertised an educational course of study, "Leadership in Aerospace." Von Ohain's reputation lured many students. His students appreciated his clear explanations and blackboard sketches, and they were fascinated that he could be distracted to reminisce about problem-solving situations from his early years in Germany. One evening class remembers the evening when von Ohain was writing a formula across the entire blackboard in the front of the room but stopped, stock still, for long minutes as he stared at it. The class worried that something had happened to him when he did not move. All of a sudden a smile crossed his face. "So that's how that works!" he said. "I was looking for that." He also was frequently quoted with saying "Oopsy!" when he would make a mistake on the blackboard.

The winter months in Dayton are gray and cold, and an opportunity was extended for Hans to teach propulsion and thermodynamics at the University of Florida at Gainesville during winter sessions of 1981–1983. Hans and Hanny visited their time-share in Florida, and their close friends Mo and Alice Lawson, who had moved there permanently. Soon the von Ohains purchased a small house where they could spend part of the year away from Dayton.

Hans and Hanny made frequent trips to visit their children. Steve and his family lived in Mississippi. Chris had married Pam in 1977 and was in California, and Cathy had married Charlie Pravel in 1979 in California. Stephanie was in college in Virginia. Because Hans was requested to speak at many seminars and received honors, he and Hanny traveled around the United States and met important people. They attended the inaugural ball for President and Mrs. Reagan and accepted the "Web of Space" trophy from John Glenn. In 1983 Hans was invited to present at a seminar in Taiwan, and they enjoyed a week of first-class hospitality in the Far East. They made almost yearly trips to Europe, several times for Hans to receive honors and also for vacations and visits with friends.

Ben Koff, former executive vice president of engineering and technology at Pratt & Whitney Aircraft Engines, told of large meetings held for up to 3000 participants where he asked Whittle, von Ohain, and the two together, to speak. He felt that everyone should know these living pioneers of history. He also brought them back to Pratt & Whitney to speak on an informal basis with the Pratt engineers and to answer questions.

Von Ohain liked walking through the Pratt factory with Koff. Von Ohain told Koff, "You are more than a vice president, you are a designer!" Koff could show him details, and von Ohain liked details in configuration, such as why there was only one turbine in the core of an engine instead of more, why there were two bearing cores instead of three, and how they had increased the tip speed. A story is told that on one tour of the factory, von Ohain stopped to ask questions of one of the workers at his workbench. The man related what he was doing, explaining that he was having some problems. The tour of VIPs stopped while von Ohain helped the workman with suggestions and got down on the floor on his hands and knees to show him where an improvement might be made.

One evening following a technical program, the scientists and engineers from Pratt had dinner at the PGA Sheraton in West Palm Beach. Koff recalls that at a later professional society dinner in Orlando, Hanny von Ohain played the piano for Sir Frank and Lady Whittle, Hans, and all the guests to sing together. Both Sir Frank and Hans enjoyed this unplanned social event.

Koff and his wife traveled in Germany with the von Ohains to attend an industry/government meeting and the Berlin Air Show. A celebratory dinner hosted by *Motoren- und Turbinen-Union* (MTU Engine Company) was held one evening in the courtyard of the

Pergamon Museum in east Berlin (this was after the Wall came down). Around them were antique treasures taken from Turkey that were property of the museum. It was said that Hitler loved to show off these acquisitions at dinner parties to impress foreign dignitaries. The evening ended with fireworks. The next day, at a ceremony in a palace in Potsdam, von Ohain was presented with the prestigious Prandtl Ring.

Koff said Hans did not enjoy being in Berlin and said there were "too many memories." He did not want to visit the house where he lived as a teenager because it reminded him too much of his family and the difficult years of the war.

Hans liked seminars as an opportunity to teach and modestly accepted special commendations, but he never lost his practical attitude. On the day of Hans's induction into the Engineering Hall of Fame at the Dayton Engineer's Club, he taught his old friend Sig Harder a lesson.

> Just as clothes were for Hans not to be shown off but to keep him warm or cool, shoes were not supposed to shine but keep him dry. When we were waiting for the limousine to take us from his house to the clubhouse, I suddenly realized that my shoes needed a shine. At that point Hans looked at them and said, "Yes, they do, and mine do, too." And then he introduced me to his "shoe shine of fame" (or shame, depending on how you feel about shoes). He lifted his right foot up and brushed it firmly five or ten times against the back of his left leg, and then he did the same on the other side. He told me proudly that was the way he had always done it, unless Hanny had other ideas, and that's the way I should do it. I have done it that way ever since.

In later reminiscences Harder and Hanny discussed Hans's modesty and total lack of pretense, and the fact that he never bought any of his clothes himself. Harder said, "Hanny suddenly covered her face, shook her head, and lowered her voice so that nobody could possibly hear her whisper, 'His shoes!'"

At home the von Ohains entertained often. If you were asked to be a guest of them in those years, you would quickly accept. Their house was in a spacious suburb surrounded by a large lawn. When you rang the doorbell the door would be opened at once by your host, with Rex, a large German shepherd, sitting by his side. Hans would draw you into the house. A man was greeted with a bow and a woman with a kiss on the hand. You would immediately be taken to a comfortable chair in the living room. A woman would have a pillow tucked beside her for comfort. A glass of wine would be placed in your hand and the host would settle down in a chair nearby to talk with you, genuinely interested in you. Hanny would appear at once with a smile. As you looked around the room you would realize you were among many interesting people, some with awesome credentials and some merely joyous good company. You would feel at home for an evening of a good meal crowded around their large table and good conversation, always with laughter and many light remarks from your host.

In 1985, von Ohain was selected as Charles Lindbergh Professor at the National Air and Space Museum of the Smithsonian Institution in Washington, DC. For a year, his assignment was to review documents and write articles for their publications. Because of his contract with UDRI he spent only three weeks of the month at the Smithsonian and the other week at UDRI.

It was a fortuitous opportunity for his daughter Stephanie, who, in seeking her fortune in the world after college graduation, wanted to go to Washington, DC, to work. She was skeptical about giving up the freedom and independence she had had at college, but she agreed to share an apartment with her father. When she found her first job, it was in the same building complex as her father's office. He gave her total freedom to live her own life, but they had a comfortable friendship of activities together. She said, "I got to know him in ways I never had at home. He was a wonderful teacher about life. I watched him at work, and saw his dignity and humanity. From his associates I learned about his great sense of humor. He had great joy in life itself." Together the "dynamic duo," as their Washington associates dubbed them, investigated restaurants of many different cultures or had dinner from a deli. Hanny, back in Dayton, became worried about Hans's cholesterol and shipped Stephanie a box of cookbooks. Hans and Stephanie puzzled over them briefly. Stephanie encouraged her father to eat a vegetarian diet during their nightly dinings-out, and he felt better.

It was not long after he arrived in Washington that he began winning friends all around him. The parking lot attendant, who liked the candy and other edible surprises he received from von Ohain, met him each morning with a smiling greeting for "Dr. V," a nickname that amused von Ohain. Von Ohain counseled Stephanie to always remember the people behind the scenes, those that really make the world go around. Von Ohain always made a point of learning the names of mechanics and receptionists, waitresses and repairmen, and had genuine interest in their lives.

Von Ohain asked one of the young men in an office down the hall to explain membership in "the coffee club." The young man teased him, asking what he had done that was important. Hans replied, "Oh, nothing," and produced a package of coffee as his contribution. When he asked the young man to show him how to make coffee with their automatic coffee pot, the young fellow shook his head in amazement that he was showing the man who invented the jet engine how to measure out scoops of coffee.

The cleaning woman for the office explained to Stephanie one afternoon that von Ohain was the nicest man who had ever worked as the Lindbergh Professor. She said she could tell because the plants in the office grew so well, and she felt plants could tell a person's goodness. What the woman didn't know was that von Ohain distributed the remainder of his cup of coffee at the end of each day into the plants. (Not recommended here.)

While Hans was working as the Lindbergh Professor, Hanny took the opportunity for a trip of her own with the Dayton Music Club for an extensive tour to Switzerland, Austria, Hungary, and Germany, visiting cultural sights and musical landmarks. The next year she went to Norway and Sweden, interrupting that tour to join Hans in Aachen where he was presented an award.

In his work with UDRI, von Ohain made contacts with scientists around the world. One was very special to him, a young Swede, Erik Prisell, who was consumed with interest in jet propulsion. Prisell had studied the work of both pioneers, Whittle and von Ohain. After a time with the Swedish Air Force and at university, he was employed at the Swedish Air Ministry as a program manager for propulsion and power systems for the Swedish fighter project, JAS 39 Gripen. An engineer who had worked with Sir Frank Whittle arranged for Prisell to meet Whittle. The same engineer encouraged Prisell to visit the von Ohains while he was in the United States on business. When Prisell called to ask when it would be convenient to visit, he was invited to stay at the von Ohain home in Dayton. He returned for visits several times, but he did not meet any of the children. In 1985 when Hanny was in Sweden on vacation she visited Erik. The story with Erik does not end there.

When Erik came to Washington from Sweden, he telephoned von Ohain at the Smithsonian to ask if they could have dinner together. One evening Hans told Stephanie that he wouldn't ask her to join them for dinner because the man he was meeting was not anyone she would know, and she would probably be bored with their company because they were going to discuss jet engines. She learned later that his dining companion was Erik Prisell and said she would have liked to meet him because Papa had talked about him for years.

When Hans returned to Dayton, Stephanie stayed in Washington, DC, to work at National Geographic. One weekend in 1988 when she was home in Dayton visiting her parents, she answered a phone call and it was Erik Prisell from Sweden. Stephanie carefully did not acknowledge that she knew all about him, but gave a very mature, "Yes?" He said he was calling to get her phone number in Washington, DC, because he had a book to give to her father, but he also hoped she would accompany him to a Swedish Embassy party in Washington the following Monday. They met and began to see each other each time he came to Washington. Erik Prisell and Stephanie von Ohain were married in 1991 and now live in Sweden with their son Kristofer. As a representative of the Swedish government, Erik invited Hans to speak at a seminar in that country, and in early 1992 Hans and Hanny got to visit their daughter's new home in Sweden.

In September 1986 the deputy for propulsion at Wright–Patterson Air Force Base celebrated its 10th anniversary with a historical symposium at Wright–Patterson and invited noted speakers. Von Ohain and Anselm Franz spoke on the German turbojet development, and speakers from Allison, Pratt & Whitney, and General Electric described their work. A published volume provides a short history of the propulsion office and the text of the speeches. The topics included the contribution of the piston engine to turbojet technology, the beginnings of jet propulsion, the Rolls-Royce V/STOL engine, and a broad look at other jet engine developments. By this time, von Ohain had given many speeches about his development of the jet engine. He tried to take a fresh look at his experiences each time, but he particularly enjoyed other speakers who might be on the same program. He renewed his interest in V/STOL technology and began an intensive study of the current literature on the subject. At 75 years old, he plunged into an analysis of possibilities.

Hans and Hanny visited their children and grandchildren in California at least once a year. Every time they went, the grandchildren, Chris and Pam von Ohain's children, Beth, Stephanie, and Hans; and Cathy and Charlie Pravel's children, Ehren and Ryan, loved to have them visit. Chris and Cathy talked often of how good it would be to come back to Ohio to live closer to their parents and to have their children in Ohio schools. On one trip Hans took the hydrofoil ski to California to work on improvements with Chris. Hans and Hanny, in their mid-70s, drove across the United States with a "surfboard" strapped to the top of the car.

On another driving trip across the country, Hans had been invited to speak at several universities, which would provide a few days of rest from driving along the way. Hans wrote the following letter on 20 October 1986 to his old associate, Wilhelm Gundermann, after their return:

> Am now back over a week from a long, wonderful trip through America. Hanny and I left with our auto at the beginning of September. We drove through the Rocky Mountains, through the fantastic canyons (Bryce and Zion Canyons in Utah, Grand Canyon in Arizona) to San Diego on the Pacific Ocean where our children live.
>
> Then we visited friends in Phoenix, Arizona and finally we went to Tucson, Arizona where I had been invited by the University of Arizona to hold a seminar. From there we drove to Colorado Springs where I presented a lecture at the invitation of the Air Force Academy and had a lively discussion with the students. Finally we drove through the first snow to Kansas, then through Missouri and Indiana back to Dayton, Ohio. Our trip was truly beautiful and rich in variety. All in all we put 6,400 miles or 10,300 kilometers behind us.
>
> During our trip through the canyons we encountered many, many German tourists. There was so much German spoken in the restaurants that one could believe he was somewhere in Germany. The tourists belonged to various tour groups. They came in wonderful large buses with all the comforts, like washing facilities, toilets, and air conditioning. Many were older [and in poorer shape than you].

His letter continued after 5 November 1986: "I received an invitation to take part in a conference of several days' length in Detroit on automobile gas turbines. ...It will probably be several years before we have an inexpensive and economical gas turbine for automobiles."

Hans was surprised that automobiles continued to be built with antiquated power plants when he felt the future should be with hybrid engines with electric power for city driving and turbine power for the country. He blamed the entrenched system for lack of progress for achieving fuel-efficient cars.

In August 1989, the Dayton–Cincinnati section of AIAA hosted an international celebration at Dayton, Ohio, for the Golden Anniversary of Jet-Powered Flight. Its goals were:

> To honor those whose vision and creativity made the jet engine a reality.
>
> A retrospective on the development of the jet engine over the past 50 years.
>
> A brief review of the current state of jet engine technology.
>
> A look to the future of the jet engine as we enter a new era of powered flight.

Attendees came from all areas of the United States. International guests included Karl Ernst Heinkel, son of Ernst Heinkel of the original *Heinkel Flugzeugwerke,* and representatives from the *Motoren- und Turbinen-Union* (MTU) from Germany, the Royal Aerospace Establishment and Royal Air Force from Great Britain, and the Swedish Air Ministry. After an all-day session, each attendee received a commemorative document with text of the speeches and illustrations. From a review of the day written by Dale Whitford of UDRI, chairman of the event, comes this:

> In addition to speeches by the two inventors, Whittle and von Ohain, the symposium program included six other speakers who have been, or are now, leaders in the field. First we heard early jet developers, Martin Hemsworth of General Electric and Harold Stetson of Pratt & Whitney, who told how their companies entered the jet age. Next were current leaders. Brian Rowe of GE presented concern about the United States' ability to develop jet engines in the future because of a projected engineering and scientific manpower shortage by 1995, as well as a challenge to industry to solve the problem. Barry Waldman of Rocketdyne described hypersonic jet propulsion and, more specifically, the NASP program. Finally, two current leaders, Alan Smith of Rolls-Royce and Bernard Koff of Pratt & Whitney

presented their projection of the next 50 years of jet propulsion in Europe and the United States, respectively.

A formal dinner completed the long day. Entertainment was performed by a musical combo and four singers. Men in black wore white vests embroidered with gold stars, and the women singers sang and danced in gold lamé jumpsuits. Their songs included, "Those Magnificent Men in their Flying Machines," "Fly Me to the Moon," and "I'm Flying" (from *Peter Pan*), with word changes to include von Ohain and Whittle. Film clips were shown of the flight of an Me 262 jet fighter past a formation of B-17s during World War II, an F-16 takeoff with afterburner, and the test of an engine with vectored thrust capability. This was followed by a pictorial review of the lives and careers of the coinventors of the turbojet engine. Awards were made to both and the evening ended with all guests singing "America the Beautiful."

Hans and Hanny had a little house in Florida during the winter months and lived in Dayton the rest of the year. They visited their children or had them come home, and they socialized with friends. Hans continued with his own projects and ideas. In the summer of 1992 they spent two weeks in Germany for Hans to be awarded the Diesel Award and Prandtl Ring. Hans was in the best of health.

In August of that year he became tired and lethargic, and his doctor suggested a blood test. He was diagnosed with endocarditis. During a spinal tap, a blood clot developed and he became paralyzed from the waist down. He spent three months in the hospital and had to have rehabilitation to learn to walk again. Stephanie flew home from Sweden to help her mother and be with her father at the hospital. When the doctor suggested that a warmer climate would be better for him, Hans and Hanny decided to sell their home in Dayton and move to Florida permanently.

Ironically, both Chris and his wife, Pam, and Cathy and her husband, Charlie, long separated from their parents across the country, made their decision at this same time to move back to Ohio from California.

With regrets, Hans and Hanny went to Florida, knowing it would be the best situation for Hans's health. Their next five years were pleasant. In Melbourne on the east coast of Florida they built a larger home with a swimming pool. Hans bought Hanny a large organ and she played for his entertainment and volunteered to play at a local senior citizens center. They invited friends and former work associates to stay with them in their pleasant home on the lake, to have breakfast on the porch, and to take drives to the ocean, just a few minutes away, or to Cape Canaveral for sightseeing. They missed their old friend Mo Lawson, who had died in 1992, but Hanny saw her friend Alice Lawson frequently.

Their son Chris was an independent consultant working in Cincinnati and was able to come to Florida frequently to work with his father on several iterations of a new design for a more efficient and faster human-powered watercraft and improvements to the hydrofoil ski. Between 1993 and 1997, Chris was able to be in Melbourne 177 days with his father. Von Ohain was happy to have help and protested to Chris, "If I were 20 years younger, you would not be able to keep up with me!" Chris countered that it was a challenge to keep up with him as it was. Although his father was methodical in his work, he worked efficiently and steadily. They worked to perfect the design and construction of the human-powered boat, building it in the garage and testing it in the swimming pool. Chris and his father grew very close to each other over these last few years and became best friends, working together against merciless time as von Ohain grew more frail. A patent for the design of the human-powered watercraft was issued in July 1998 after Hans von Ohain's death.

In his last few years Hans received a constant flow of requests through the mail for verifications and recommendations about his work. A friend suggested to him that he should think more of himself and free himself from the cumbersome burden of answering all the hundreds of letters when he was no longer well. His friend said, "You have to say 'No.' You are too kind!" Hans replied, "Too kind? I did not think that that was possible."

Requests for speeches in many locations continued. Ben Koff remembers the last time Hans came to Pratt & Whitney. Hanny drove him there from their home in Melbourne. Hans was already very ill, more ill than anyone realized. Hans did not want to disappoint the people who came to hear him, but he was too weak to make a speech. At the last minute, the two worked an arrangement that Koff would field a question and pass it to Hans who would answer it, and Koff would repeat the answer. For an hour and a half, this "put-up show" went on, and everyone was delighted with the format, especially considering that Hans's German accent was sometimes difficult to understand. No one realized that this was the only way the presentation could take place.

In the fall of 1997, Hans developed severe chest pains, which were diagnosed as congestive heart failure. He had to stay in rehabilitation centers for more than 100 days. He rallied in the daytime when Hanny visited, calling her "the sunshine of my day." He came home to be limited to a large easy chair or wheelchair during the day. By February of 1998, after a hospital stay, he was sent home to be cared for by hospice nurses who had to sedate him with morphine for his pain. As their son Steve grieved, "No man should ever have to suffer as he suffered." The children gathered at the von Ohain home in Melbourne, Florida. Hans had enough strength to reach up to grasp their hands and say to each, "I love you." Hans died on 13 March 1998, surrounded by his adored wife and children.

At the time von Ohain retired from Wright–Patterson, he said he had an idea that would revolutionize the Air Force, and he wanted to have time to work out the details. Sadly, the idea will forever remain a mystery. We will not know if his project was a jet engine with no moving parts, his original concept of an engine from his student days. He made a sketch of this idea late in his life, but admitted that he needed to do more work on it. Perhaps he would have solved the difficulties with V/STOL aircraft. After his retirement he researched current literature on the subject in depth, and in 1988 he joined the American Helicopter Society to receive their literature. It may have been another idea that he never expressed. Whatever the invention, he did not finish it. He left it for another mind to achieve.

The man who wanted "elegance" for flight according to both definitions, "refinement and grace of style" and "scientific exactness and precision," found what he wanted in the turbojet engine he invented. He also demonstrated both definitions in how he lived his life.

Chapter 20

Memorials

Hans von Ohain's friends say his qualities were generosity, tolerance, willingness to help, natural kindliness, courage, and optimism for the "leap into the unknown." They say, "In a practical way, he was untraditional, unconventional, and unafraid of failure."

There were occasions for praise. Rudolf Friedrich, who worked with von Ohain in the 1940s at the Heinkel-Hirth Company, made this statement at the award of the Aachen and Munich Prize for Technology and Applied Science in 1985:

> To know Hans von Ohain is to have a glimpse of the life of a genius, full and rich in every regard, a man rich in ideas, self-reliant, clear of thought, and with the power to inspire.
>
> The main features of his character, which impressed me so when he was 28 years old, are still there. He spoke openly of his problems and ideas, sought and accepted criticism, recognized good suggestions and the performance of others without envy. He remained free of pride and arrogance. He knows and openly acknowledges that he was first because he concentrated on the simplest solution, and Heinkel's stormy urging helped him overcome every obstacle.

Brian Quinn, who worked with von Ohain at the Aerospace Research Laboratory, wrote a technical paper titled "HvO's First Legacy," published in a U.S. Air Force Wright Aeronautical Laboratories (AFWAL) *Collection of Papers in the Aerospace Sciences* in 1982. This is the introduction:

> To have contributed to a volume paying homage to HvO's research and development career in the Air Force honors this author. To have experienced the explosion of ideas, the questioning smile and the curiosity of Dr. von Ohain through daily

association is to have learned to extract the essential, to have tasted success in science and to have shared his humanity. And, hopefully, to have appreciated the friendship of an especially exciting human being.

As comfortable with Hegel as with Heisenberg, finding parallels between minor themes in Antigone and loss mechanisms in ejectors, acquaintances may see Dr. von Ohain as a man of contrasts. His friends, on the other hand, readily discern the continuous threads which unify such contrasts within an intellect of vast proportions.

So it is that the man who in reviewing a mathematical development once said, "At my age, Brian, one no longer integrates," was also the man who years before had so exquisitely integrated not just a mathematical expression but several sciences and technologies into the first jet engine. That engine flew in 1939.

The commander of the Air Force Aero Propulsion Laboratory at the time of von Ohain's retirement in 1979 was Col. (later Brig. Gen.) Philippe O. Bouchard. On 3 April 1998, he spoke at a memorial service for von Ohain at Grace Methodist Church, Dayton, Ohio.

History already has permanently recorded Hans von Ohain's distinguished contributions to technology and accorded to Hans his place of honor amongst the world's greatest inventors. I share with you some of my fond recollections of Hans and, in doing so, I hope to represent the thoughts and affection of his many Air Force friends

It was my good fortune to be the commander of the Air Force Aero Propulsion Laboratory at Wright–Patterson Air Force Base in the late 1970s. Hans, a relative newcomer to the lab, was the chief scientist. He had been assigned to this position to establish and oversee an in-house research program in an organization which heretofore had a contract-out culture. He rose to the challenge admirably by nurturing, with great patience, the lab's junior scientists and engineers as they ventured into hands-on research in an atmosphere of considerable skepticism. Only a person with Hans's reputation and understanding of sound research could have achieved the success which today is a thriving research program which undergirds all propulsion exploratory and advanced development in the directorate.

Part of his responsibility was to select a few high-risk, high-payoff projects which would compete for funding at the highest levels in Washington. The projects were called "laboratory director funded projects," and, as such, had to be briefed by me. Hans patiently spent many hours with me, tutoring on thermodynamic complexities well beyond my understanding, to assure our success in obtaining the required funding.

Patience with his commander, patience with our junior scientists and engineers, but impatience with bureaucracy! Paperwork and red tape were an anathema to him and provoked him to use strange German expressions which I never asked him to translate. He conducted a perpetual but friendly war with Frank Patella, our director of operations, concerning the lab's copying machine which almost always broke down when Hans used it. He claimed his first demonstration jet engine was more reliable.

Humility was another of his attributes. We both tried to outdo ourselves in showing deference for one another. He, on the one hand, would bow and call me "my commander." I, on the other hand, tried to show my reverence for Hans, realizing full well that neither he nor I nor the lab, for that matter, would have been there, had it not been for his invention. His humility was evident in the manner in which he gave other people a large share of the credit for his accomplishments.

If ever there was a "civil scientist" in our lab, certainly it was gentleman Hans. Ever gracious in his dealings with people, he was frequently heard saying, "Wonderful, wonderful, wonderful!" I loved to hear his robust laughter from his office or the lab hallways as he joked with one and all. If our resident jet engine inventor was upbeat, I told myself, things could not be as desperate as they might seem to me.

Hans's continuing creativity and inventiveness are legendary. This creativity was not constrained to the invention of the jet engine, as his 19 U.S. patents attest. Hans had more creative energy in any given year than most of us have in a lifetime.

You might think that an internationally recognized person such as Hans might be a bit standoffish with his fellow human beings. Not so! His lab office door was always open, eagerly welcoming one and all for congenial, intellectually stimulating discussion. If not in his office, he wandered the lab's hallways and project sites, showing genuine affection for all he saw.

> Above all, he loved his family, the top priority in his life. At the 1978 "Encounter of the Jet Engine Pioneers," Hans and Sir Frank [Whittle] were asked about compensation received for the jet engine invention. Whittle replied, with Hans wholeheartedly agreeing: "Neither of us made much money out of the jet engine invention, but both of us had excellent taste in blonde wives." Both had their priorities right!

Hans von Ohain is described well in Gen. Bouchard's final words: "This patient, humble, gentle, creative, family-oriented person just happened to have invented the jet engine."

Appendix A

Patented inventions by von Ohain

German Patents

Hans von Ohain was awarded more than 50 patents in Germany between 1935 and 1945 during his employment with Ernst Heinkel. He lost all rights to these inventions with Germany's defeat in World War II.

U.S. Patents at Aerospace Research Laboratory

[Patent numbers have seven digits, applications for patents have six, inventions identified for a patent have four numbers. Information is incomplete.]

Date	Patent No./Serial No./Invention No.	Description
1956	Application # 177,632	Injector Pump (with M. O. Lawson)
Nov. 1959	2,911,189	Fluid Machine
Jan. 1964	3,119,233	Multiple Electrode Arrangement for Producing a Diffused Electrical Discharge (with Frank L. Wattendorf and M.O. Lawson)
March 1966	3,217,488	Gas Cooled Colloid Propulsion Systems (with M. O. Lawson and S. Hasinger)
Oct. 1966	3,225,225	High Voltage Electrostatic Generator (with Frank L. Wattendorf, M. O. Lawson, and S. Hasinger)
Aug. 1968	3,379,011	Energy Exchange Device and Separator (with Roscoe H. Mills)
June 1969	3,439,197	Generation of Ions in a High Pressure High Velocity Gas Stream (with M.O. Lawson and F.L. Wattendorf)

Sept. 1969	3,465,180	Two Component Electrofluid Dynamic Powered Generator Employing Contact Ionization (with Frank L. Wattendorf)
Jan. 1970	5673	Electrofluid Dynamic Power Generator
Oct. 1970	3,535,850	Centrifugal Particle Separator (with Roscoe H. Mills)
Nov. 1970	3,525,474	Jet Pump or Thrust Augmentor (with Charles A. Scolatti and Roscoe H. Mills)
Feb. 1971	6804	Electrofluid Dynamic Process and Devices for Direct Conversion of Enthalpy into Electromagnetic Radiation (with Maurice O. Lawson)
Sept. 1971	3,573,512	Electrofluid Dynamic Generator System (with Frank L. Wattendorf, M.O. Lawson, and J.A. Decaire)
Oct. 1971	7913	Jet Wing with Multiple Thrust Augmentors (with Col. Charles Scolatti)
Feb. 1972	3,618,322	Cavity Reactor with External Separator (with Melvin R. Keller)
Feb. 1972	7917	Reentry Heating Energies Analyzer (with Elmer Johnson)
Jan. 1973	3,711,370	Colloidal-Gas Core Reactor (with Melvin R. Keller)
April 1973	3,713,970	Cavity Reactor with Two Stage Separation (with Melvin R. Keller)
July 1973	3,724,784	Wing with Thrust Augmentor (with William S. Campbell)
Nov. 1973	3,739,634	Apparatus for Generating Ultra High Total Enthalpy Gases with Multicomponent Flow (with Elmer Johnson)
April 1974	3,785,593	Aircraft Propulsion System (with C.A. Scolatti and Roscoe H. Mills)
Sept. 1976	3,981,144	Dual Stage Supersonic Diffuser (with Col. Robert W. Milling)
May 1975	3,770,227	Jet Wing with Multiple Thrust Augmentors (with C.A. Scolatti)
Sept. 1975	3,893,335	Device for Simulating Reentry Conditions in a Particle-Laden Atmosphere (with Elmer Johnson)

U.S. Patents at the University of Dayton Research Institute

Date	Patent No./Serial No./Invention No.	Description
March 1982	4,321,006	Gas Compression Cycle and Apparatus Therefor (with Frank L. Wattendorf and M.O. Lawson)
April 1984	4,439,988	Rankine Cycle Ejector Augmented Turbine Engine (with John E. Minardi, M.O. Lawson, and Ival O. Salyer)
Sept. 1987	4,689,950	Hot Gas Flow Generator with No Moving Parts (with John E. Minardi)
July 1988	4,756,154	Hot Gas Flow Generator with No Moving Parts (with John E. Minardi)
Dec. 1988	4,791,784	Internal Bypass Gas Turbine Engines with Blade Cooling (with John E. Minardi and M.O. Lawson)

U.S. Patent (Independently Received)

Date	Patent No./Serial No./Invention No.	Description
July 1998	5,785,564	Water Craft with a Hydrofoil-Bladed Wheel Assembly (with Christopher H. von Ohain)

Appendix B

Honors received by von Ohain

Late in his career Hans von Ohain received commendations not only for what he had done on the turbojet engine in Germany, but for the work he was leading and producing for the benefit of aerospace research in his position with the U.S. Air Force.

Government Awards

Outstanding Performance, including a citation for sustained superior performance, 1964. He continued to receive Outstanding Performance awards from the Air Force in the following years, an unusual achievement because in the year following an award more is expected of the individual than in the year previous.

Special Act or Service from both Aeronautical Systems Division (ASD), the parent organization of the Aerospace Research Laboratory (ARL), and from the Office of Aeronautical Research (OAR) for the development of supersonic axial compressors, 1964.

Decoration of Exceptional Civilian Service, the highest award the Department of the Air Force can bestow on a civilian, 14 June 1965. The justification stated, "Since 1955, Hans von Ohain has contributed significantly to the defense capabilities of the United States through his comprehensive analysis of needs and potentialities in the area of aerospace propulsion."

Air Force Engineering Achievement Award, 1966. For the introduction of overhead concepts to make it possible to use smaller gas core nuclear reactors for propulsion.

Air Force Meritorious Service Award, 1967. Citing his technical contributions.

Eugene M. Zuckert Management Award, 1967. For his ability to work effectively with both government and industry personnel. The justification stated:

> The Eugene M. Zuckert Management award is presented yearly to recognize top-level Air Force managers. The award is

named after Eugene M. Zuckert, who served as Assistant Secretary of the Air Force from 1947 until 1952 and Secretary of the Air Force from 1961 to 1965.

Secretary Zuckert was a staunch advocate of teaming between the military and civilian workforce to capitalize on the military's knowledge of warfighting and the continuity provided by civilians.

The Zuckert Management award is a mahogany plaque bearing the Air Force coat of arms and individual silver plates with the name of each winner. Recipients keep the award for one year until passing it to the next winner. Honorees also receive a citation and two-inch silver-plated medallion bearing the Air Force seal.

Air Force Systems Command nominee for National Civil Service League Career Service, 1974.

Air Force Association Citation of Honor naming him Air Force Civilian of the Year, 1978. The justification stated: "Outstanding leadership, creativity, technical competence and foresight are among the exceptional traits Dr. von Ohain has brought to his job as chief scientist. His tireless efforts have resulted in advanced propulsion concepts vital to the national defense. His contributions have brought national and international acclaim."

Department of Defense Distinguished Civilian Service Award, 1 August 1979. For exceptional devotion to duty and extremely significant contributions to Defense Department operation. The justification stated: "He has given unsparingly of his time and talents in service of the Department of Defense, which reflects credit upon himself and civilian employees of the DoD."

National and International Awards

AIAA Goddard Award, presented jointly to Hans von Ohain and Frank Whittle by AIAA, 25 January 1966. The justification stated:

> The award, named in honor of rocket pioneer Robert H. Goddard, is given on a national basis to a person who has made a brilliant discovery or a series of outstanding contributions over a period of time in the engineering science of propulsion or energy conversion.
>
> For outstanding scientific creativity and leadership in accomplishing the first successful application of turbojet propulsion to aircraft and for continued excellence in planning and conduct-

ing pioneering research in the engineering sciences of propulsion and energy conversion.

AIAA, Fellow, from 1970.

Jet Pioneers' Association of the U.S.A., Distinguished Member, 1974.

Wright Brothers Memorial Trophy at Andrews Air Force Base, 1976.

Wings Club Distinguished Achievement Award, New York, 1980. (Note: At this presentation a replica He S 3B engine was on display. It was one of two built by the Deutsches Museum, Munich, from drawings re-created by von Ohain. The original engine was destroyed in the bombing of Berlin during World War II. This reproduction was transferred to the National Air and Space Museum of the Smithsonian Institution, while the other was retained at the Deutsches Museum.)

National Academy of Engineering, Member, 1980.

International Aerospace Hall of Fame, San Diego, California, 1982

Engineering and Science Hall of Fame, Dayton, Ohio, 1985.

Web of Space Award, Washington DC, 1985.

Aachen and Munich Prize for Technology and Applied Science *(Aachener und Munchener Preis fur Technik und angewandte Naturwissenschaften),* 1985.

International Society for Airbreathing Engines (ISOABE), 1987

Munich Celebration of 50 Years of Flight, 1989.

Golden Cross *(Grosses Verdienst Kreutz),* Germany, 1990.

Detroit Press Club, 1990.

National Aviation Hall of Fame, 1990.

Deutsches Museum, 1991.

Wright Brothers National Memorial, First Flight Shrine Inductee, 17 December 1991.

Tom Sawyer Propulsion Award, American Society of Mechanical Engineers, 1990.

Elvie Lawrence Smith Lecture Award, Pratt & Whitney Canada, 1990.

Guggenheim Medal, sponsored jointly by ASME, SAE, and AIAA, 1991. The justification stated: "For pioneering development of turbojet propulsion resulting in the first flight in 1939 and lifetime achievements in aeronautical propulsion dynamics." (This award had been given to Frank Whittle in 1946 for his jet engine development. Other recipients were Orville Wright in 1929, William Boeing in 1934, Charles Lindbergh in 1953, Theodor von Kármán in 1955, Charles Stark Draper in 1966, and Gerhard Neumann in 1979.)

Charles Stark Draper Prize and gold medal, awarded jointly to Hans von Ohain and Sir Frank Whittle, 1991. This is the highest prize for science, equal to the Nobel Prize. The justification stated: "For engineering innovation and individual tenacity in developing and reduction to practice of the turbojet engine, thereby revolutionizing the world's transportation system, improving the world's economy, and transforming the relationships between nations and their peoples."

(The large gold medal von Ohain received was donated to the University of Dayton in 1993. The citation at that location says, "For engineering achievement that contributes to human welfare and freedom. One has only to look skyward to sense the magnitude of Whittle's and von Ohain's engineering in this regard.")

Rudolf Diesel Award, Germany, May 1992. Award of a gold medal. His picture hangs in the Patent Hall of German Engineers.

Prandtl Ring, Pioneer of the First Hour *(Pionier der ersten Stunde),* Potsdam, Germany, 17 June 1992. German Society for Aeronautics and Astronautics.

Honorary Doctorates

West Virginia University, University of Dayton, and Milwaukee School of Engineering.

Honorary Posts

Aeronautical Research and Development (AGARD) Advisory Group of the North Atlantic Treaty Organization (NATO), 1966–1971. Panel member representing the United States for the Propulsion and Energetics Panel.

Charles Lindbergh Professor at the National Air and Space Museum of the Smithsonian Institution, 1984–1985.

APPENDIX C

Reconstruction of the He S 3B engine

Reconstructions of Hans von Ohain's He S 3B engine that powered the first flight are on display at the National Air and Space Museum of the Smithsonian Institution in Washington, DC, and the Deutsches Museum in Munich, Germany. The original in a museum in Berlin was destroyed during a bombing in World War II. The replicas are true to the original in almost every detail, both in design and material. The engines are cut open to expose the inner workings. A visitor can set the engine in motion with a small handwheel and see firsthand how the air flowed through the compressor, combustor, and turbine.

On the following pages are reductions of the redrawn blueprints of the He S 3B engine and excerpts of some of the correspondence exchanged by the German museum and von Ohain during the reconstruction of two replicas of the engine.

The engines were rebuilt from drawings prepared under the guidance of von Ohain. In 1973 von Ohain was contacted by the Deutsches Museum concerning the construction of a full-sized model of the He S 3B jet engine for display at the museum. The Smithsonian Institution had asked him for drawings a few years earlier when they planned to build a model for the opening of a new exhibit hall in 1976, and von Ohain, at his own expense, hired a draftsman to create detailed drawings. A company recommended by the Smithsonian entered into a contract with him to build the model at no cost to von Ohain, guaranteeing him a percentage of the profits from the sale of the model to offset his expenses. The model was not built because of a reallocation of funding.

Von Ohain provided the Deutsches Museum with the seven blueprints he had prepared earlier, which, he said, "represent a fully sufficient specification for the construction of the model. Should the model builder wish a mounting diagram, I will gladly identify all subassemblies and describe how and in what order they were assembled."

In addition, he said, "I have recalculated the aerothermodynamic quantities. These calculations of the static performance and the background material make it possible to calculate the in-flight performance without difficulty."

Motoren- und Turbinen-Union (MTU) Deutsche Aerospace at Munich and numerous additional propulsion-related firms worked with the German museum to reconstruct a pair of He S 3B engines. The first replica was presented to the National Air and Space Museum on 8 December 1980 for display in its Jet Aviation section. The second was placed in the Air and Space Travel Hall of the Deutsches Museum on 15 May 1981.

Von Ohain gave the German museum addresses and background for three men in the United States who could provide the historical picture of development of jet propulsion:

> Dr. Anselm Franz, Stratford, Connecticut, who was responsible for the development of the Jumo 004 jet engine.
>
> Hans Mauch, Dayton, Ohio, who was the leader of jet propulsion development in the Technical Office of the RLM until 1939. He was, from the technical standpoint, influenced by Heinkel's turbojet development, but was convinced that this development was too important and urgent to be carried out in an airframe company. In his view, it belonged in an engine firm, which would have a larger experience base and suitable development engineers. Mauch initially had great difficulties in interesting the engine companies in the development of jet propulsion. He left the RLM at the end of 1939.
>
> Helmut Schelp, Pacific Palisades, California, who was Mauch's successor. He possessed an astounding[ly] broad view of the immense future importance of jet propulsion and a comprehensive knowledge of this kind of propulsion and its various future forms. An example is his planning reports which he presented to the *Academie der Deutschen Luftfahrt Forschung* [Academy of German Aviation Research], yearbooks 1939 and following. Schelp was not against the continuation of Heinkel's developments. Rather, he planned to develop a strong competitive program in Germany by bringing the engine manufacturers in and by distributing the various types and performance classes in the most favorable way.

Von Ohain also cited a former coworker in Germany:

> Wilhelm Gundermann, Nuremberg, who contributed much to the success of the first flight engine. He possessed an outstanding knowledge of fluid machinery and in drafting. In addition, he knew his way around Heinkel very well. He knew every section leader, and he was very familiar with the administrative processes of the firm. He eased the transition from physicist to engineer for me. As the first airplane, the He 178, was laid out, he played a decisive role in the integration of engine and airframe. He strongly influenced the basic design of the He 178. He designed an air inlet which had the greatest independence from air speed and angle of attack, and he sized the inlet and exhaust pipes for the optimum installation and minimum internal losses. On the basis of his experience in airframe and engine construction, Mr. Gundermann later took an important position in airframe construction where he was responsible for engine integration and related problems.

The following are excerpts of letters of clarification provided to the Deutsches Museum, dated if available. Various drawings and pencil sketches accompanied the comments but are not included here.

Letter, Undated:

> The development of the first flight engine stood under enormous time pressure. The mechanical design began at the end of 1937 and was supposed to be done by the middle of 1938. At the end of 1937, many problems of the internal aerodynamics and the combustion chamber were still unsolved.
>
> The requirements for the turbine bearing were carried out in cooperation with a ball- and roller-bearing firm. The following technical data were given to this firm:
>
> –Maximum axial displacement between turbine shaft and turbine bearing +/- 2 mm.
>
> –Temperature of the fuel-bearing carrier about 70 degrees C.
>
> –Temperature of the turbine shaft end about 350 degrees C.
>
> –Running time between turbine bearing inspections about five hours.
>
> Results:
>
> 1–Roller bearing with axial control of the rollers on the inner race, outer race smooth with elongated bearing surface. This

allowed the desired axial movement of the shaft relative to the bearing carrier. The disadvantage was in assembly. The outer race belonged to the bearing carrier while the inner race and rollers belonged to the rotor. Very careful assembly was necessary.

2–In consideration of the high temperature of the turbine shaft end, a bushing was fitted between the shaft and the inner bearing race as a thermal barrier. This bushing had only very small contact areas with the shaft.

3–For lubrication, a packing of hot-bearing grease was provided. It had to be refilled about every four hours.

4–Assembly was difficult and had to be carried out very carefully. For assembly, the outer bearing race belonged to the bearing carrier, as already mentioned, and the cover disk, to the rotor. A jig eased assembly.

5–The lubrication by hot-bearing grease was adequate for the demonstration flights of the He 178. It was clear, however, that this grease represented only a temporary solution. Lubrication with oil, with oil cooling the bearing and recovery of the oil, was necessary. The development of such a system required time and manpower, which were not available. This problem was not solved until 1942, with the He S 011 engine.

Letter, von Ohain to Deutsches Museum, 19 November 1976:

I had not placed any emphasis on the accuracy of the blade angles, as necessary only for an operating machine, not for a display model. If you want to build an apparatus with true blade angles in order to enable a total performance I will gladly do the calculations, taking a few days vacation at the end of December and dedicating myself to the task.

Letter, von Ohain to Deutsches Museum, 17 December 1976:

The mass flow of the He S 3B is given as 20 kg/sec in postwar reports. In my opinion this value is not correct. At an overspeed of 11,600 rpm it was only about 12 kg/sec. In order to be sure, I recalculated the jet velocity based on thermodynamic data which I remembered well. The value of the jet velocity came out to be over 400 m/sec. For a static thrust of 500 kp, that corresponds to a mass flow of only 12 kg/sec.

m = k x 500/410 = 12 kg/sec (approx)

A mass flow of 20 kg/sec is unsupportable from a thermodynamic standpoint and moreover doesn't match the dimensions of engine, ducts, and exit nozzle.

The axial precompressor should be equipped with fourteen to sixteen blades like the He S 8A and not with eight as shown in my sepia originals [previously submitted].

The structural arrangement of the combustion system was as follows: The combustion took place in an annular chamber to which air was delivered through sixteen individual 180-degree bent tubes. Each of these bent tubes had a wedge-shaped center divider with the broad end at the combustor entrance. Inside each wedge at the combustor entrance was a gas manifold from which five or six little gas distributor tubes went out. These little tubes each had four or five holes of about one mm diameter which discharged the gasoline vapor in the direction of airflow.

For ignition, we initially used a hydrogen pilot flame. The ignition of the hydrogen stream was by a single spark plug. The electrical spark was produced with a Ford Model T ignition coil, which was used instead of the usual ignition coil with an automatic impulse breaker. The starting procedure was as follows:

1) Compressed air was introduced into the starting nozzle, bringing the rotor up to about 1,200 rpm.

2) The electric ignition was switched on and the hydrogen valve was opened. The fine stream of hydrogen was instantly ignited.

3) The throttle lever was advanced to the start position, letting a small quantity of liquid fuel into the annular combustor. The liquid fuel ignited immediately from the hydrogen flame and a bright red-yellow flame filled the entire volume of the combustor, the turbine, and part of the exit nozzle. The machine accelerated, and because of the intense radiation, the gasoline vaporizing tubes were rapidly heated (10 to 15 seconds). This was recognizable through the total disappearance of the yellow flame.

4) The hydrogen was turned off and the throttle advanced to idle. The speed increased to about 6,000 rpm in about ten seconds or less. At this speed the exhaust temperature was quite low, about 360–400 degrees centigrade. The static thrust amounted to about 70 kg.

Later we replaced the hydrogen pilot with a single spark plug. The ignition proved to be just as prompt as with the hydrogen pilot. I can't remember if we employed this simplification at the time of the investigations with the He 178. I know positively that the He S 8A which was mounted in the He 280 was equipped with such an ignition.

The thrust loss from the inlet and exhaust ducts in the He 178 was as follows: the total pressure at the exit was approximately 15% lower as installed in the airplane, compared to a test stand. That meant that the exit velocity was reduced about 7%. In order that the operating point of the engine was not changed, the mass flow had to be held constant. This was accomplished by increasing the nozzle exit area about 7% so that with the same mass flow and a 7% lower exhaust velocity the static thrust loss was only 7%. In contrast to Siegfried Guenter's opinion, we always measured a static thrust loss in the installed state compared to the uninstalled engine, but by virtue of the adjustment of the exit area described above, the static thrust loss was finally reduced to 35 kg.

Letter, 9 June 1978: Von Ohain answered the following questions that had been sent on 26 April 1978 by Walter Rathjen at the Deutsches Museum. (All direct quotes.)

Question 1–Are the compressor and turbine disks really as thick as you have drawn them? Of what material are they made? Do you still know what the weight of the rotating assembly was? As drawn, and made of steel, we estimate about 220 kg, and that seems too high to us.

a. The turbine and compressor disks were basically thinner than drawn (the thick disks were planned for a hardwood model), outer thickness about 1.5 mm. Center shaft was correspondingly longer and likewise thinner.

b. Hubs and center shaft: cast steel, turned to size. Compressor blades: duralumin. Turbine blades: nickel steel. Compressor and turbine disks: steel (possible because of a cooling air shroud over the turbine disk and hub).

c. The total rotating assembly weighed about 140 kg initially. I believe we got it down to 120 kg.

Question 2–What kind of seal was used on the rims of the rotor disks?

None. The gap between disk and housing amounted to about 2 mm. The static pressure difference between the compressor and turbine gap was small, and a cold air shroud was provided on the turbine side.

Question 3–*How was the rear bearing lubricated? Since the lubrication proved to be one of the biggest problems, we would like to show it.*

The axial position of the rotor was established by the front ball bearing which also carried the axial thrust. This bearing was lubricated by pressurized oil drops; the injected oil was not recovered. Since the rotor and housing expanded differently in operation, provision had to be made for an axial displacement between the turbine end of the shaft and the turbine bearing carrier.

The turbine bearing carrier was cooled with fuel. Initially we tried to cool the turbine end of the shaft this way also. Since we could find no feasible heat exchange method, we gave up this idea, and only the central cooling duct remained with no practical function.

Question 4–*Were the shafts solid or bored out?*

The turbine and compressor hubs were not bored out in the center, but recesses were provided in both hubs.

Question 5–*Were there ribs on the housing? You have drawn a very large rib on the compressor wall but none at all on the combustor and turbine housings. How was it in reality? The MTU design provides ribs. These would have a strong influence on the external appearance. How did you otherwise finish the housing? Was it cast or turned?*

a. I believe that I remember that the compressor wall had large ribs. The reason was to maintain absolute rigidity because the gap between the compressor blades and housing was only about 1 mm. This part of the housing was made of cast aluminum.

b. The front part of the housing was likewise made of cast aluminum. The inside was turned to size. I believe the outside was ribbed as your drawing shows.

c. The turbine housing was made of cast steel (with much excess material) and was turned to shape on both sides. Therefore it had no ribs. The gap between housing and turbine blades was about 2 mm and became somewhat bigger in operation.

Letter, von Ohain to the Deutsches Museum, 31 July 1980:

> I have been trying hard to remember how the He S 3B was mounted. I really believe that we had a four-point mount. The rear part of the fuselage of the He 178 was connected to the front via four lengthwise stringers. I am absolutely certain that the aluminum skin was only a covering and carried no load. One of the stringers, the top one or the bottom one, was removable. I believe it was the bottom one. The upper stringer carried a cable channel for the flight controls.
>
> I believe the engine was installed as follows:
>
> 1) The skin and lower stringer were removed.
>
> 2) The rear part of the airplane was jacked up.
>
> 3) The engine was inserted into the fuselage from below, I believe with a slight inclination to the horizontal in order to bring the mounts clear of the stringers. If the engine were installed from above, the locations of the four mounts would have to change. Naturally, the control cable channel would then be on the bottom stringer.
>
> 4) The mounts rested on the stringers and were bolted to them. I cannot recall any discussion of thermal expansion. The steel outer housing of the engine reached a temperature of about 160 degrees C. [320 F.] (max). The resulting thermal stresses seemed to lie well in the elastic region, and slotted holes were to my knowledge not considered. As I recall, several methods of attaching the mounts to the engine housing were tried: bolts, rivets, and welding. I believe a special welding process was chosen which caused no warpage and which made the thickness of the mounts in the vicinity of the weld about 2 mm greater.

On 20 October 1980, Hans von Ohain was honored with a Distinguished Achievement Award by the WINGS Club in New York City. When he and Hanny arrived at the hotel where he was to speak, they were surprised to find a completed reproduction of the He S 3B engine that had been sent from Germany and a large display of pictures from the years he was with the Heinkel Company. The replica shown that day was taken for permanent display to the National Air and Space Museum. On 14 July 1981, von Ohain attended a ceremony opening the exhibit.

In May 1981, Hans and Hanny were in Munich as guests of Karl-Ernst Heinkel, Ernst Heinkel's son, for the celebration of the addition of the He S 3B engine to the Deutsches Museum. They toured *MTU Deutsche Aerospace,* where the replica had been faithfully rebuilt.

Drawings of the He S 3B Engine

On 22 December 1973, under the guidance of von Ohain, redrawings of plans for the He S 3B engine were completed. This engine powered the flight of the He S 178 aircraft for the world's first turbojet-powered flight on 27 August 1939. The following are reductions of these drawings.

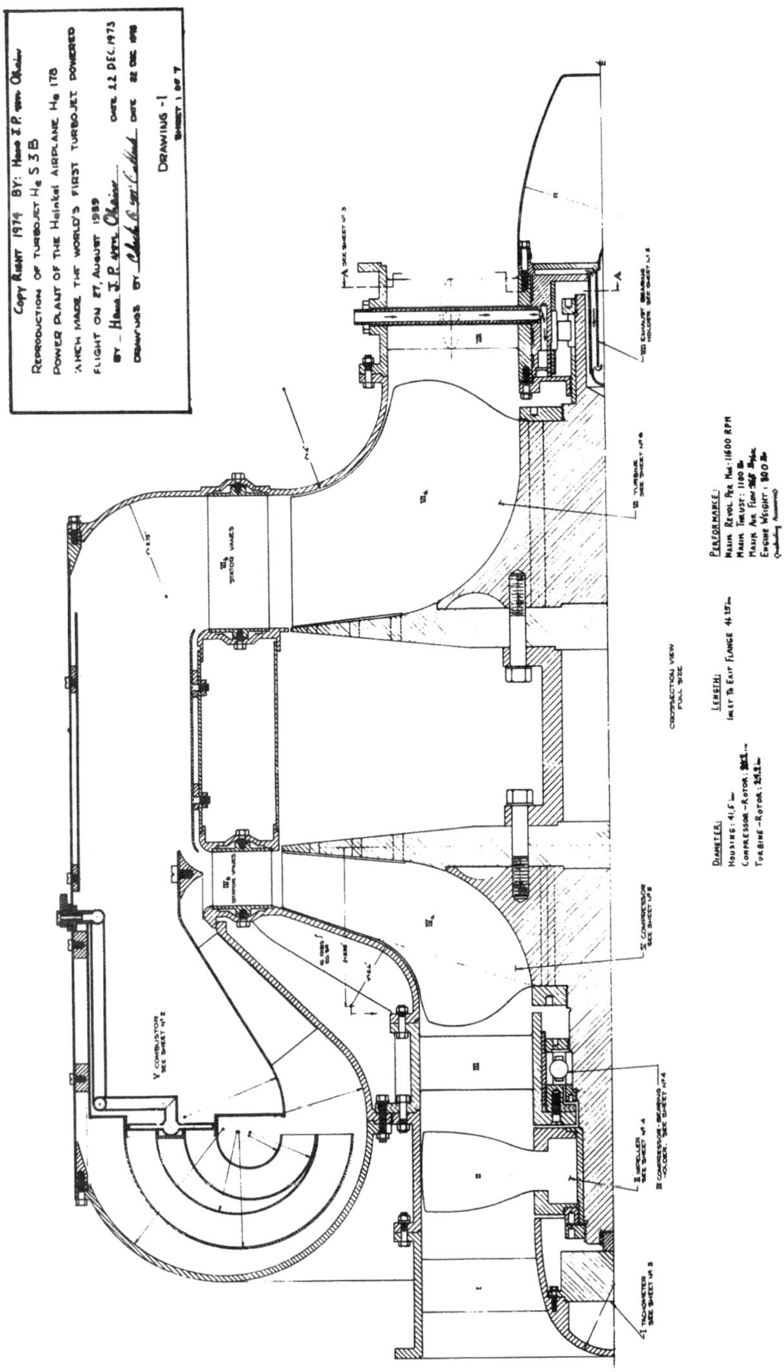

Fig. 1 Cross-section of He S 3B.

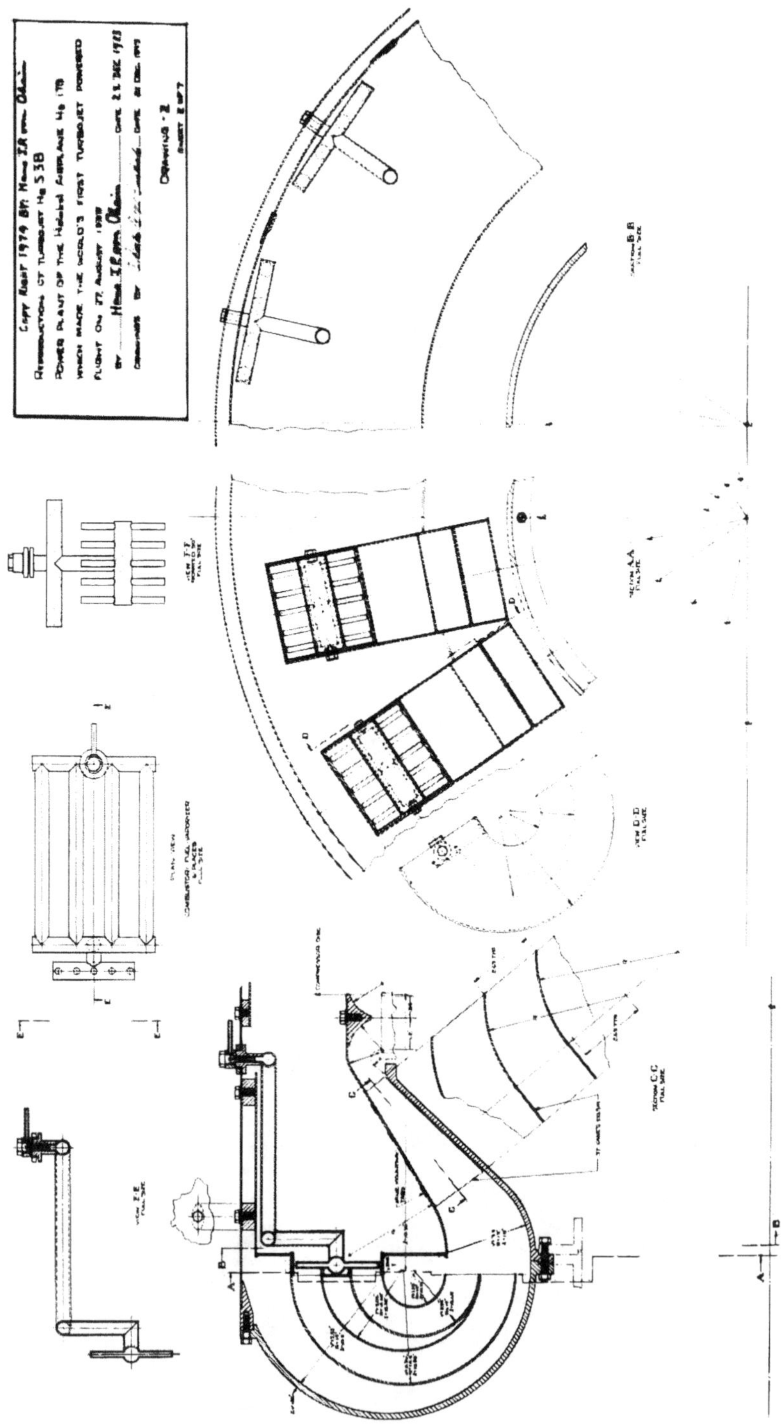

Fig. 2 Arrangement of combustor flowpath and fuel vaporizer installations.

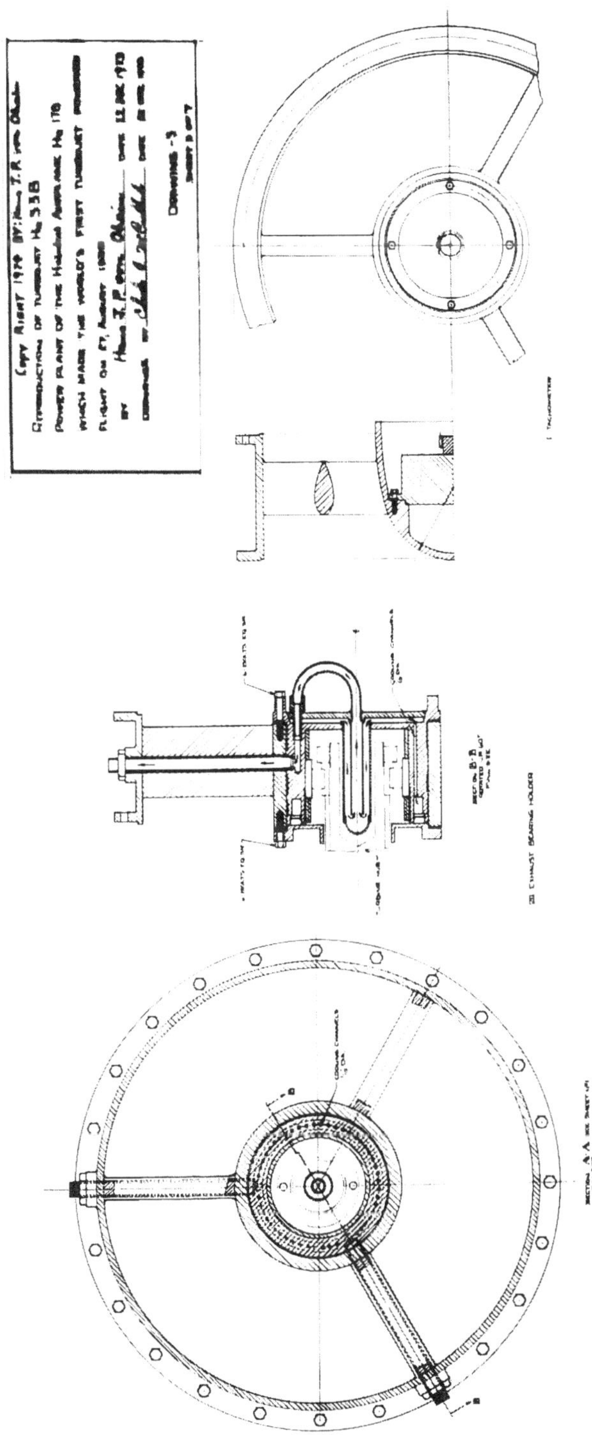

Fig. 3 Rear bearing support structure and inlet structure with tachometer installation.

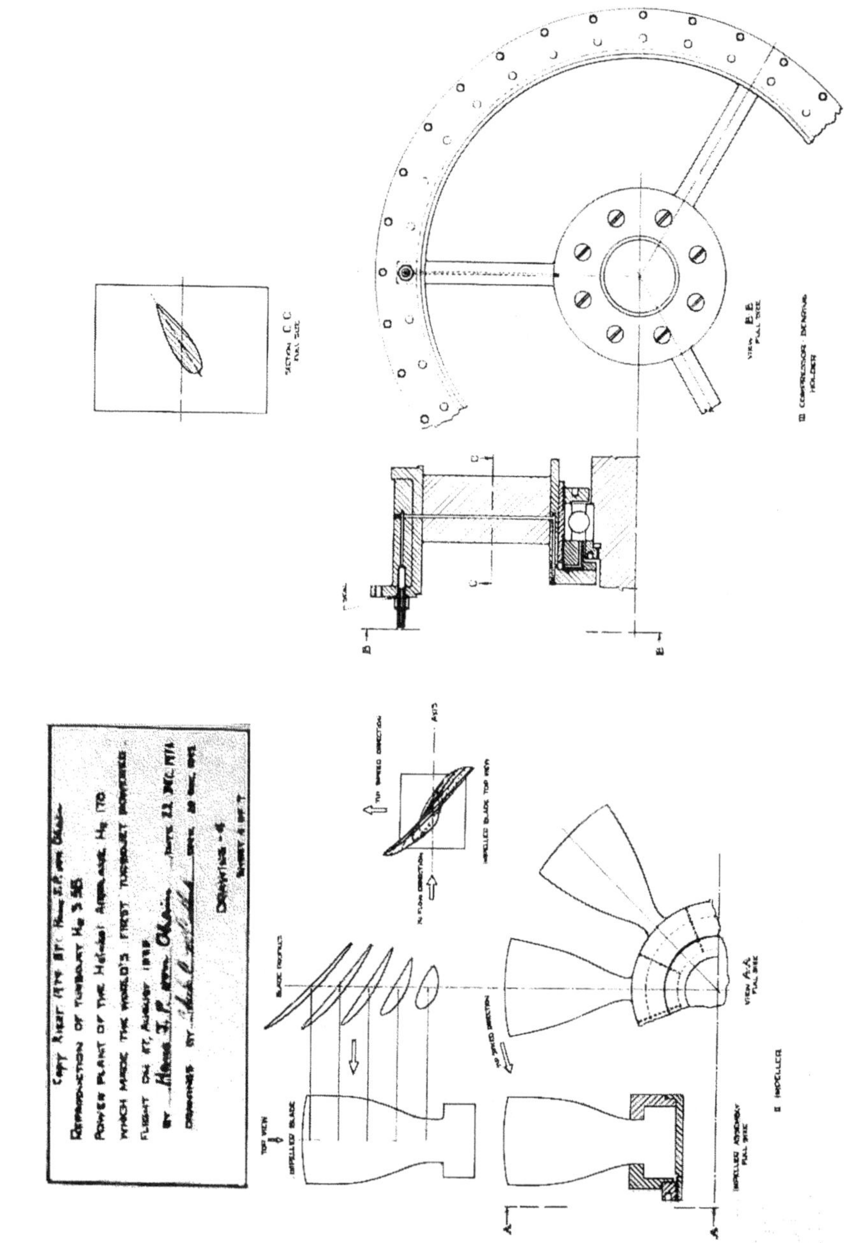

Fig. 4 Details of impeller constructions and compressor bearing installation.

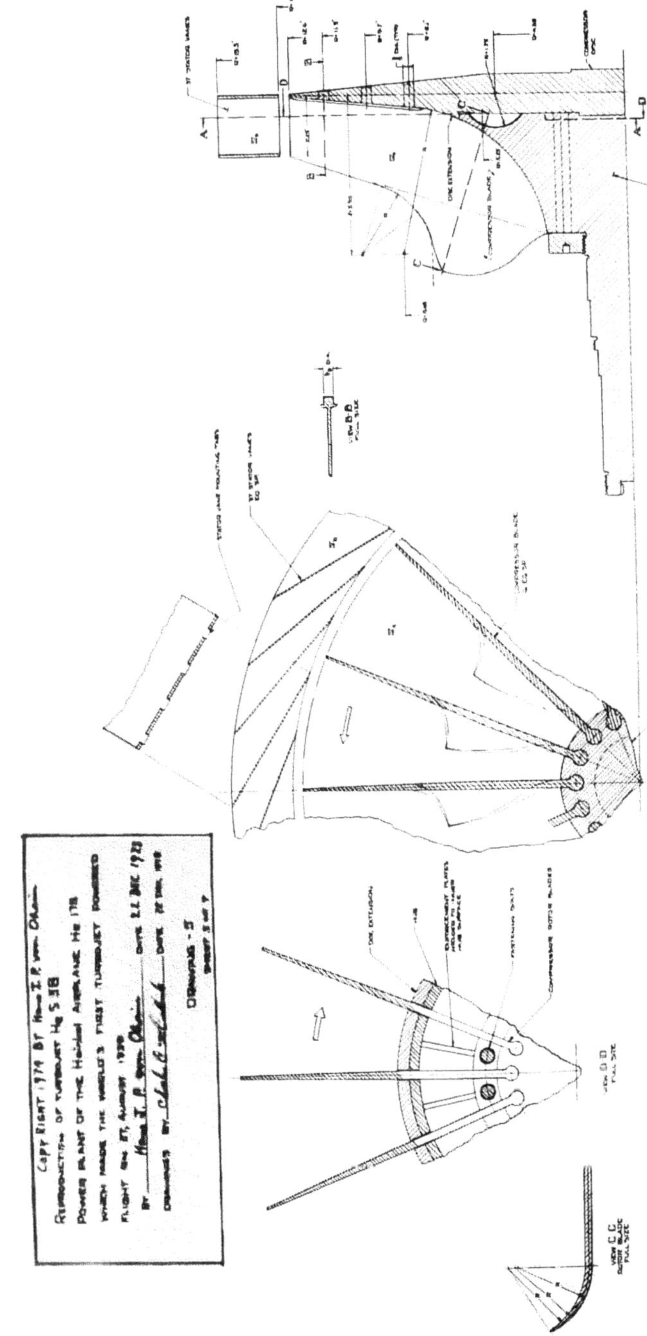

Fig. 5 Design details of compressor rotor and diffusers.

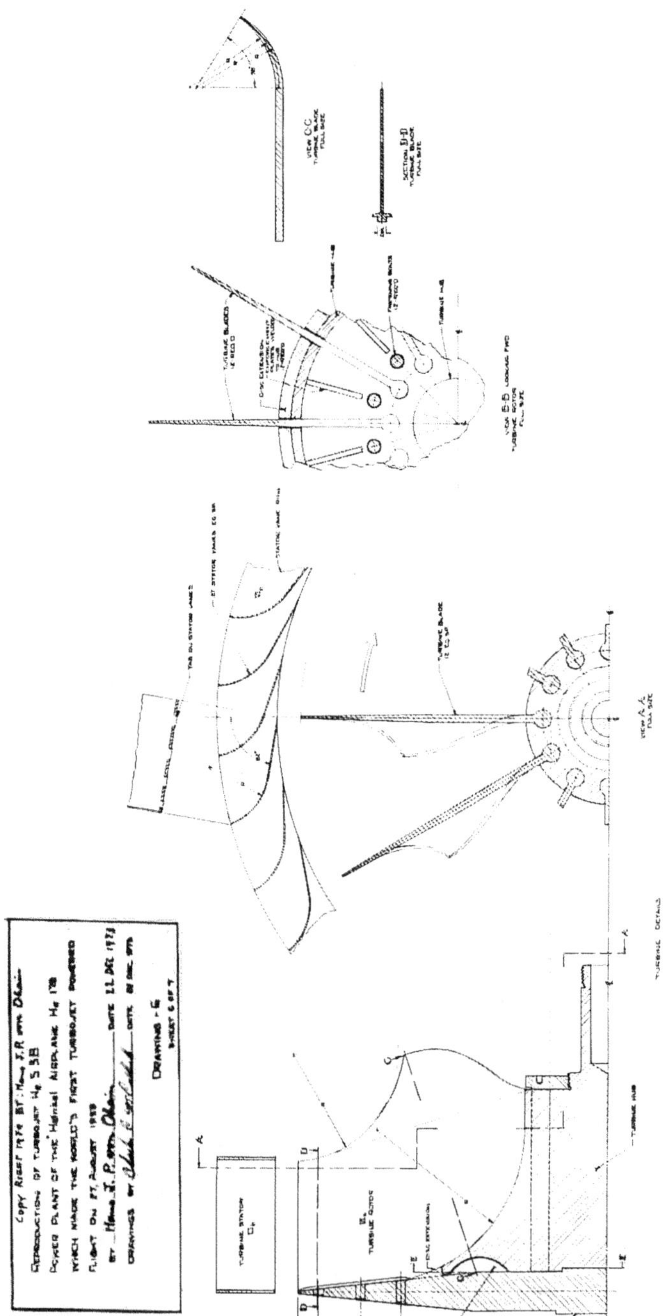

Fig. 6 Design details of turbine.

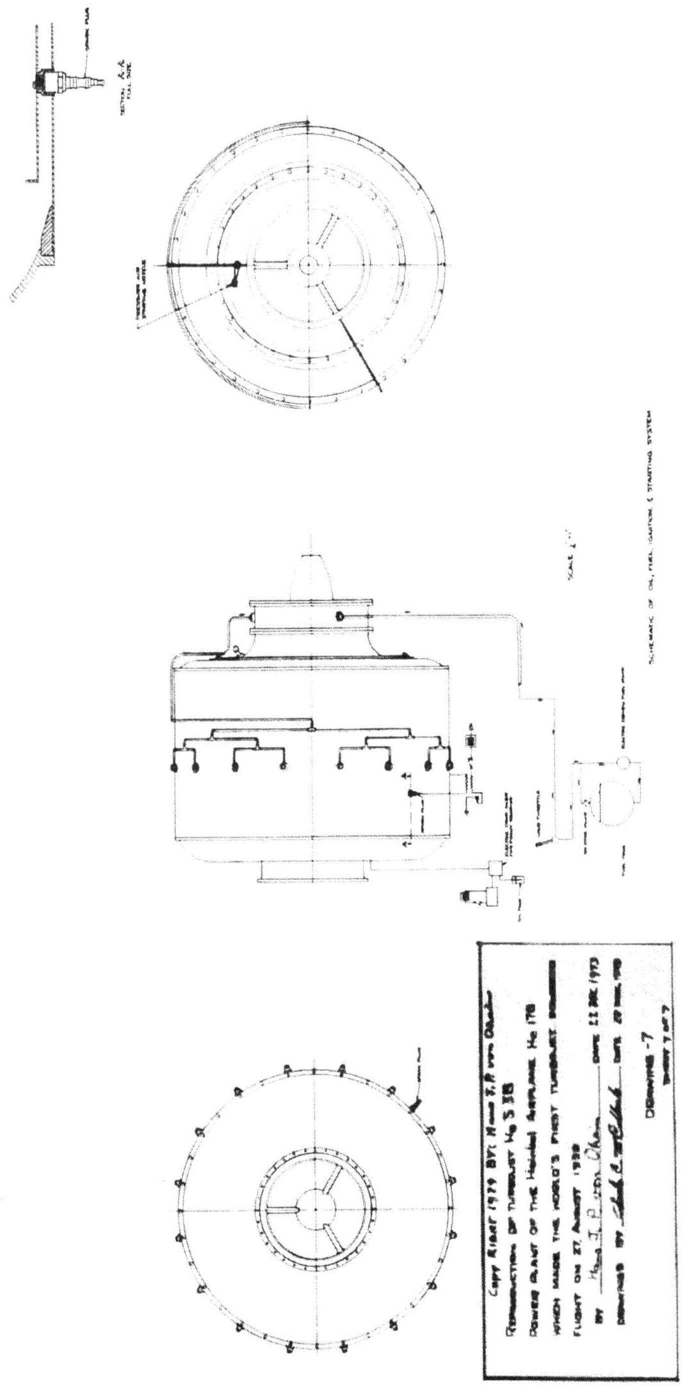

Fig. 7 Schematic arrangement of oil, fuel, ignition, and starting systems.

APPENDIX D

Data for Heinkel engines invented by von Ohain

Engine		He S3B	He S8A	He S011
Year		1939	1941	1944
Thrust	(lbs)	989	1124	2940
	(N)	4400	5000	13100
Fuel Consumpt.SFC(lbs/lbfh)		1.6	1.82	1.35
	(kg/Nh)	0.163	0.186	0.137
Thrust/frontal area (lbf/sq ft)		105	218	537
	(N/m2)	5043	10417	25739
Airflow	(lbs/s)	26	33	64
	(kg/s)	12	15	29
Pressure ratio		2.8	2.7	4.2
Turbine Inlet temperature (F)		1270	1290	1427
	(C)	690	700	775
RPM		13000	13500	10200
Compressor		1a*)+1r	1a*)+1r	1a*)+1d+3a
Combustor		Annular	Annular	Annular
Turbine		1r	1r	2a
Length	(ft)	3.44	5.2	10.97
	(m)	1.048	1.58	3.34
Diameter	(ft)	3.46	2.56	2.64
	(m)	1.054	0.782	0.805
Weight	(lbs)	793	838	1951
	(kg)	360	380	885

a= axial; r = radial; d = diagonal or mixed flow
*) Inducer

Chart created by Erik Prisell, January 2000.

APPENDIX E

Timeline for German jet engine development

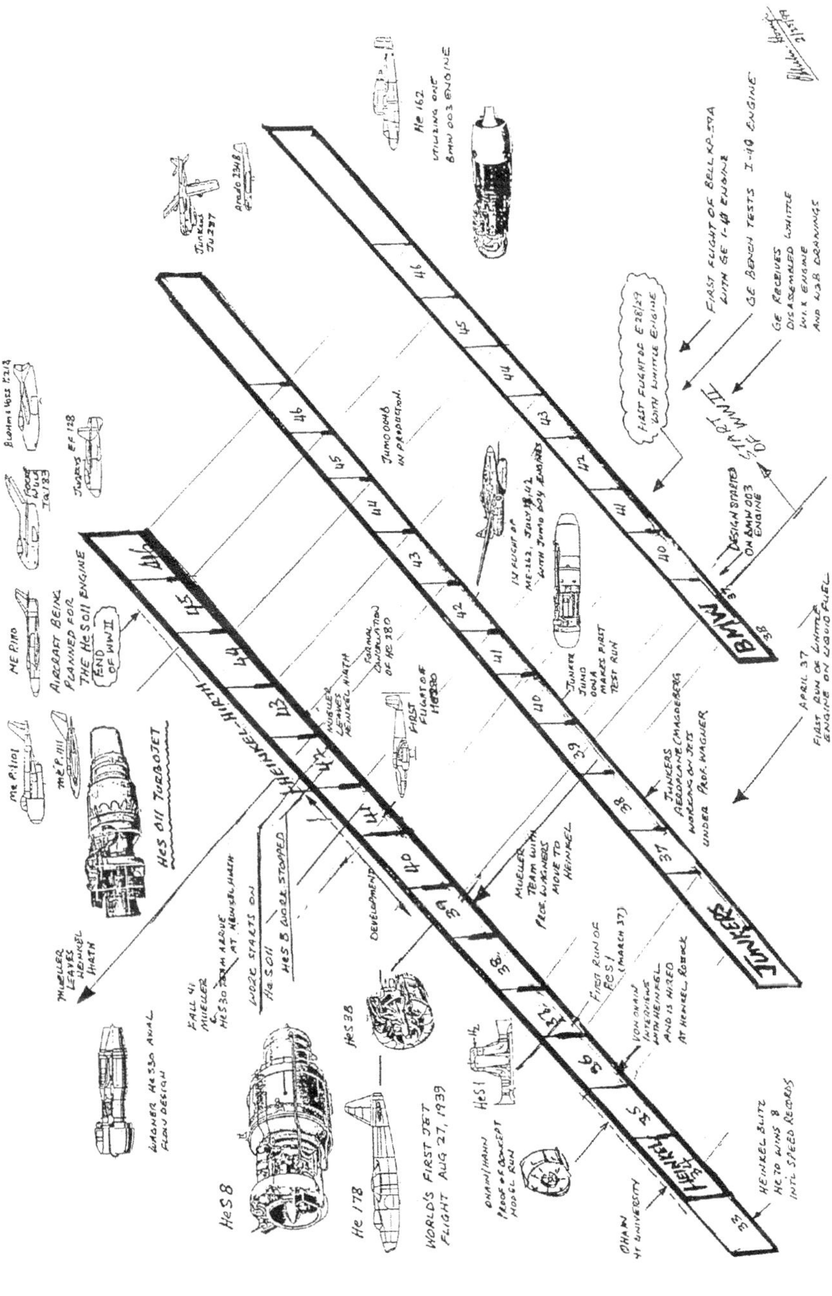

Timeline showing German jet engine development program and operational and planned turbojet aircraft. Numbers in squares indicate years. (Chart created by Cyrus Meher-Homji.)

Resources

Written by von Ohain

"The Evolution and Future of Aeropropulsion Systems," *The Jet Age,* edited by Walter Boyne and Donald Lopez, National Air and Space Museum, Smithsonian Institution Press, Washington, DC, 1979, pp. 25–46.

"The Origins and Future Possibilities of Air Breathing Jet Propulsion Systems," Eighth International Symposium of Air Breathing Engines, AIAA, 1987.

"Aircraft Engines–Part I," *Air World,* Vol. 40, No. 2, 1988, pp. 8–9.

"The Beginnings of Jet Propulsion," Deputy for Propulsion's 10th Anniversary Historical Symposium, Wright–Patterson AFB, OH, 4 September 1986; published by HQ Aeronautical Systems Division (AFSC) History Office.

"The First Turbojet Flights and Other German Developments," Celebration of the Golden Anniversary of Jet Powered Flight, 1939–1989, Dayton OH, 23 August 1989, published by AIAA.

Foreword, *Elements of Gas Turbine Propulsion,* by Jack D. Mattingly, McGraw-Hill Inc., New York, 1996, pp. xv–xiii.

Chapter Twelve, "Jet Development Under Heinkel and von Ohain," *Casting About,* No. 1, Howmet Corp., Greenwich, CT, 1990, pp. 8–13.

"The Past, Present, and Future of the Gas Turbine Engine," National Air and Space Museum Research Report, Smithsonian Institution Press, Washington, DC, 1985, pp. 45–66.

Written Record of Conversation between Whittle and von Ohain

An Encounter Between the Jet Engine Inventors, Sir Frank Whittle and Dr. Hans von Ohain, History Office, Aeronautical Systems Division, AFSC, Wright–Patterson AFB, OH, 3–4 May 1978.

Transcriptions of Formal Interviews

Hans von Ohain with J. J. Ermenc of Thayer School of Engineering, Dartmouth College, Hanover, NH, Wright–Patterson AFB, OH, 29 October 1964.

Hans von Ohain with James St. Peter, included in Ernest C. Simpson interviews as "HvO, former associate of Ernest C. Simpson," 16 and 23 January 1986.

Books Used for Reference

American Heritage Dictionary, Second College Edition, Houghton-Mifflin Co., Boston, MA, 1982.

Amtmann, Hans A., *The Vanishing Paperclips: America's Aerospace Secret Monogram,* Aviation Publications, Boylston, MA, 1988.

Bar–Zohar, Michel, *The Hunt for German Scientists,* Hawthorn Books, New York, 1967.

Bentele, Max, *Engine Revolutions: The Autobiography of Dr. Max Bentele,* Society of Automotive Engineers, Warrendale, PA, 1983.

Bower, Tom, *The Paperclip Conspiracy,* Little Brown & Co., Boston, MA, 1973.

Boyne, Walter, *Clash of Wings, World War II in the Air,* Simon & Schuster, New York, 1997.

Boyne, Walter, and Lopez, Donald (eds.), *The Jet Age: Forty Years of Jet Aviation,* National Air and Space Museum, Smithsonian Institution Press, Washington, DC, 1979.

Constant, Edward W., II, *The Origins of the Turbojet Revolution,* The Johns Hopkins Univ. Press, Baltimore, MD, 1980.

Eimerl, Sarel, *Hitler Over Europe, the Road to World War II,* Little, Brown & Co., Boston, MA, 1968.

Fulbrook, Mary, *A Concise History of Germany,* Cambridge Univ. Press, Cambridge, MA, 1990.

Gay, Peter, *My German Question,* Yale Univ. Press, New Haven, CT, 1998.

Gordon, Lois, and Gordon, Alan, *American Almanac, Year by Year Through the 20th Century,* Yale Univ. Press, New Haven, CT, 1999.

Green, William, *Warplanes of the Third Reich,* Doubleday, Garden City, New York, 1970.

Griehl, Manfred, *German Jets of World War II, Warbirds Illustrated,* Arms and Armour Press, London, 1974.

Hallion, Richard P., *The Epic of Flight, Designers and Test Pilots,* Time-Life Books, Alexandria, VA, 1984.

Haug, C. James (ed.), *The Strategic Defense Initiative, An International Perspective,* Columbia Univ. Press, New York; Social Science Monographs, Boulder, CO; Center for Internal Security and Strategic Studies, Mississippi State Univ. Strategic Studies Series, Vol. 1, 1987.

Heimann, Grover, *Jet Pioneers,* Duell, Sloan, and Pearce, New York, 1963.

Heinkel, Ernst, *Stormy Life: Memoirs of a Pioneer of the Air Age,* E. P. Dutton & Co., New York, 1956.

Hogg, Ian V., *German Secret Weapons of the Second World War,* Greenhill Books, Lionel Leventhal, London, 1999.

Hunt, Linda, *Secret Agenda: The United States Government, Nazi Scientists, and Project Paperclip, 1945–1990,* St. Martin's Press, New York, 1991.

Irving, David, *The Rise and Fall of the Luftwaffe: The Life of Field Marshall Erhard Milch,* Little, Brown and Co., Boston, MA, 1973.

Jarausch, Konrad H., *The Unfree Professions: German Lawyers, Teachers, and Engineers, 1900–1950,* Oxford Univ. Press, New York, 1975.

Josephy, Alvin M., Jr., (ed.), *American Heritage History of Flight,* American Heritage Publishing Co., Simon and Schuster, Upper Saddle River, NJ, 1962.

Jones, Glyn, *The Jet Pioneers, the Birth of Jet-Powered Flight,* Metheun London, Michelin House, 1989.

Kee, Robert, *In the Shadow of War,* Little, Brown & Co., Boston, MA, 1939.

Killen, John A., *History of the Luftwaffe, 1915–1945,* Berkley Publishing, Garden City, NY, 1969.

Koehler, H. Dieter, *Ernst Heinkel—Pionier der Schnellflugzeuge,* Bernard & Graefe, Koblenz, Germany.

Knappe, Seigfried, and Brusaw, Ted, *Soldat, Reflections of a German Soldier, 1936–1949,* Orion Books, New York, 1992.

Lasby, Clarence G., *Project Paperclip: German Scientists and the Cold War,* Atheneum, New York, 1971.

Mitcham, Samuel W., *Eagles of the Third Reich: The Men Who Made the Luftwaffe,* Presidio Press, Novato, CA, 1970.

Mason, Herbert Molloy, Jr., *The Rise of the Luftwaffe: Forging the Secret German Air Weapon, 1918–1940,* Ballentine Books, New York, 1976.

Murray, Williamson, *The Luftwaffe 1933–1945, Strategy for Defeat,* Air University Press, Maxwell AFB, AL; Superintendent of

Documents, U.S. Government Printing Office, Washington, DC, 1983.

Neville, Leslie E, and Silsbee, Nathaniel, *Jet Propulsion Progress,* McGraw-Hill, New York, 1948.

Overy, R. J., *The Air War 1939–1945,* Scarborough House, 1981.

Pletschacher, Peter, *Heinkel; Chronik und Typenblatter der Firma Heinkel-Flugzeugbau,* Aviatic Verlag, Munich, 1989.

Parkinson, Roger, *The Origins of World War II,* G. P. Putnam Sons, New York, 1970.

Schlaifer, Robert, *Development of Aircraft Engines,* Division of Research, Graduate School of Business Administration, Harvard Univ., Boston, MA, 1950.

Sears, Stephen W., *Air War Against Hitler's Germany,* American Heritage Publishing Co., New York, 1963.

Shirer, William, *The Rise and Fall of the Third Reich: A History of Nazi Germany,* Simon and Schuster, New York, 1959.

Smith, G. Goeffrey, *Gas Turbines and Jet Propulsion for Aircraft,* Dorset House, London (and earlier versions), 1977.

Smith, J.R., and Kay, Antony, *German Aircraft of the Second World War,* Putnam & Co., London, 1972.

Speer, Albert, *Inside the Third Reich,* Macmillan Co., New York, 1970.

St. Peter, James, The History of Gas Turbine Engine Development in the United States—A Tradition of Excellence, International Gas Turbine Institute of the American Society of Mechanical Engineers, Atlanta, GA, 1999.

St. Peter, James (ed.), *The Memoirs of Ernest C. Simpson, Aero Propulsion Pioneer,* Aero Propulsion Laboratory, Air Force Wright Aeronautical Laboratories, Wright–Patterson AFB, OH, 1987.

Steinhoff, Johannes, Pechel, Peter, and Showalter, Dennis (eds.), *Voices from the Third Reich, an Oral History,* Regnery Gateway, Washington, DC, 1989.

Sunderman, James F. (ed.), *World War II in the Air—Europe,* Franklin Watts, New York, 1964.

Vincenti, Walter G., *What Engineers Know and How They Know It,* Johns Hopkins Univ. Press, Baltimore, MD, 1993.

Von Kármán, Theodor, *The Wind and Beyond,* Little, Brown & Co. (Canada), 1971.

Wagner, Wolfgang, *The First Jet Aircraft,* Schiffer Publishing, Atglen, PA, 1998.

White, Michael, and Gribben, John, *Einstein: A Life in Science,* Dutton, New York, 1994.

Whiting, Charles and editors of Time/Life Books, *The Home Front: Germany,* Time/Life Books, 1982.

Other Documents

Friedrich, Rudolf, *Making the Aircraft Gas Turbine a Reality,* translation of address given at Technische Universität Karlsruhe, Germany, undated.

Gundermann, Wilhelm, *Twenty Years of Turbojet Flight,* translation of address given 27 Aug. 1979.

Nagaraja, K.S., and Petty, J.S. (eds.), *A Collection of Papers in the Aerospace Sciences,* AFAPL-TR-79-2126, AFAPL, AFWAL, AFSC, Wright–Patterson AFB, OH, 1982.

Prisell, Erik, *In Memory of Hans von Ohain 1911–1998,* written at time of his death, March 1998.

Schubert, Helmut (ed.), *50 Jahre Turbostrahlflug, 50 Years of Jet-Powered Flight,* transcription of speeches given at the symposium of *Deutsche Gesellschaft für Luft- und Raumfahrt e.V.* (German Society for Air and Space Travel), Munich, 26–27 Oct. 1989.

Viets, Hermann, "Hans von Ohain, Contributions of Excellence," part of *To Honour Hans von Ohain,* written for his retirement, 1979.

Walker, Lois E., and Wickam, Shelby E., *From Huffman Prairie to the Moon, A History of Wright–Patterson AFB,* Air Force Logistics Command, available from Superintendent of Documents, U.S. Government Printing Office, Washington, DC, undated.

Articles about von Ohain in Magazines and Newspapers

Boyne, Walter J., "The Evolution of Jet Fighters: A New Point of View," *Air University Review,* Jan.–Feb. 1984, pp. 35–45.

Brown, Capt. Eric, "Mastering Heinkel's Minimus," *Air Enthusiast,* June 1972, pp. 295–300.

Cauley, Kathleen A., "Jet Propulsion: Hans von Ohain Had a Dream and He Made it Come True," *Dimensions,* Univ. of Dayton Alumnus, June 1980, pp. 4–5..

Czaplyski, Vincent, "Profile: Hans von Ohain," *AOPA Pilot,* June 1992, pp. T1–T6.

Gaffney, Tim, "He Launched the Jet Age," Dayton Journal-Herald, 22 May 1984.

Hart, John, "Hans von Ohain: An Opposite Reaction," *New Zealand Wings,* Aug. 1989, pp. 30–32.

Jefferson, Patricia, "Hans von Ohain," *AIAA Student Journal,* Vol. 25, No. 3, 1987, pp. 2–6.

Johnson, Brian, "The First of the Jets," *Aeroplane Monthly,* July 1992, pp, 48–51.

Jones, Jack, "Builder of First Successful Jet Now Trying for Nuclear Rocket," *Dayton Daily News,* 23 Aug. 1970, p. 24–A.

Kennedy, Richard L., "The Jubilee of Jet Flight—1939–1989," *Leading Edge,* Winter 1989–1990, pp. 17–21.

Kreye, Andrian, "The Man Who Invented the Jet: Hans Joachim Pabst von Ohain," *Lufthansa Bordbuch,* 4 March 1990, pp. 22–28.

Rizvi, Teri, "Remember This Guy When You're Circling LaGuardia," *Dayton Magazine,* Jan.–Feb. 1990, pp. 18–19.

Rizvi, Teri, "Jet Flight: Higher, Faster, Smoother," *Dimensions,* Univ. of Dayton Alumnus, Vol. 8, No. 1, Autumn 1989, p. 1.

Schaefer, Fritz, "I Flew the World's First Jet Fighter," *Royal Air Force Flying Review,* Vol. XIV, Nov. 1958, pp. 23–27.

Scott, Phil, "Birth of the Jet Engine," *Mechanical Engineering,* Vol. 66–71 117, No. 1, Jan. 1995, pp. 66–71.

Turk, Paul, "Fascination: One of Two 'Fathers' of Jet Propulsion Alive and Well in Dayton, Thank You," *Dayton Journal-Herald,* 4 Aug. 1975.

Vietmeyer, Noel, "Created the Jet Age," *Readers Digest,* May 1987, pp. 163–166.

Vonada, Damaine, "The Man Who Changed the World," *Dayton Magazine,* March–April 1981, pp. 19–21, 41.

Vonada, Damaine, "The Man Who Would Go Faster," *Ohio Magazine,* March 1989, pp. 39–41, 63.

____"Face to Face with Hans Pabst von Ohain," *Aerospace America,* Vol. 28, No. 7, July 1990, pp. 8–9.

____"Development of the German Jet Warplane," *Air War,* Vol. 1, 1986, pp. 4–13, 82–83.

____"When the First Turbojet Aircraft Flew," *Aviation,* Oct.–Nov. 1978, pp. 36–41.

____*"Wie das Jet-Zeitalter in Göttingen begann"* (How the Jet Age Began in Göttingen) and *"Die Liebe zur reinen Physik, zur leisen Fliegerei zum Leben"* (The Love of Pure Physics, of Quiet Flight, of Life), *Göttingen Magazin,* 30 Dec. 1994.

____"Meine Liebe war die reine Physik" (My Love was Pure Physics), Universität Göttingen *Spektrum,* Feb. 1995.

Original Documents

Hanny von Ohain.

Deutsches Museum.

Archives of the Heinkel Company.

History Office of Aeronautical Systems Center, Wright–Patterson AFB, OH.

History Office of Air Force Aero Propulsion Laboratory, Wright–Patterson AFB, OH.

Archives of the National Air and Space Museum of the Smithsonian Institution, Washington, DC.

Index